# THE
# PSYCHOLOGY
# OF LEARNING
# MATHEMATICS

# THE
# PSYCHOLOGY
# OF LEARNING
# MATHEMATICS

Expanded American Edition

## Richard R. Skemp
Emeritus Professor, University of Warwick

LAWRENCE ERLBAUM ASSOCIATES, PUBLISHERS
1987   Hillsdale, New Jersey                    Hove and London

Lawrence Erlbaum Associates, Inc., Publishers
365 Broadway
Hillsdale, New Jersey 07642

*Library of Congress Cataloging-in-Publication Data*

Skemp, Richard R.
    The Psychology of learning mathematics.

    Bibliography: p.
    Includes index.
    1. Mathematics—Study and teaching.   I. Title.
[DNLM:   1. Mathematics—education.   2. Psychology,
Educational.   QA 11 S627p]
QA11.S57      1987        370.15′6      87–6767
ISBN 0-89859-837-0
ISBN 0-8058-0058-1 (pbk.)

Printed in the United States of America
10  9  8  7  6  5  4  3  2  1

# Contents

**PART B**

# Notes and Acknowledgments

In Part A, chapters 2–7 are the same as in the original edition (Skemp, 1971) except for small revisions in the wording. Chapter 1 has been entirely rewritten for the present edition.

Part B of the original edition showed how the ideas of the first part could be applied to the understanding of various topics in mathematics, beginning with conceptual analyses of some of the most basic ideas. In the present edition, this has been replaced by new material representing further developments in my thinking.

Chapters 8, 9, 10, 11, and 16 are newly written for the present edition (as also was chapter 1). The remaining chapters first appeared as journal articles. I am grateful to Mr. Dick Tahta, editor of *Mathematics Teaching,* for agreeing to the reproduction here of the articles that form chapters 12, 13, 15, and 18. I am grateful to Dr. Merald E. Wrolstad, editor of *Visible Language,* for agreeing to the reproduction of the article that now appears as chapter 14. And I am grateful to Dr. Thomas Romberg for agreeing to the incorporation in chapter 10 of a substantial amount of material from a chapter that I contributed to *Addition and Subtraction: A Cognitive Perspective* (Lawrence Erlbaum Associates, New Jersey), which he co-edited with Dr. Thomas Carpenter and Dr. James Moser; and from a longer paper given at the Wingspread Conference Center in Racine, Wisconsin, November 1979.

# THE
# PSYCHOLOGY
# OF LEARNING
# MATHEMATICS

# PART A

# Introduction and Overview

1

## WE STILL HAVE A PROBLEM

A distinguished American mathematician, who is Past President of the International Commission on Mathematical Instruction and who has written and spoken extensively in recent years on the teaching of mathematics to young children, recently began an invited address to another international body (Whitney, 1985) as follows:

> For several decades we have been seeing increasing failure in school mathematics education, in spite of intensive efforts in many directions to improve matters. It should be very clear that we are missing something fundamental about the schooling process. But we do not even seem to be sincerely interested in this; we push for 'excellence' without regard for causes of failure or side effects of interventions; we try to cure symptoms in place of finding the underlying disease, and we focus on the passing of tests instead of meaningful goals. (p. 123)

The situation is similar in the United Kingdom. Since the early 1960s, there have been intensive efforts to improve mathematical education in our schools, by intelligent, hard-working, and well-funded persons. Nevertheless, a research group based at London University recently reported that many children still understand some of the most important topics in mathematics no better after two or three years of secondary schooling than when they entered (Hart, 1981).

Not only do we fail to teach children mathematics, but we teach many of them to dislike it. Concern at the governmental level about the state of mathematical education in our schools led in 1978 to the formation of a committee of enquiry, whose meetings continued over a period of three years. The first stage of the inquiry consisted of interviews with people chosen to represent a stratified sample of the population. One of the most striking things about these interviews is those which did not take place.

> Both direct and indirect approaches were tried, the word 'mathematics' was replaced by 'arithmetic' or 'everyday use of numbers' but it was clear that the reason for people's refusal to be interviewed was simply that the subject was mathematics . . . half of the people approached as being appropriate for inclusion in the sample refused to take part. (Cockcroft, 1982, p. 6)

Such had been the result, for them, of ten years of so-called mathematical education while at school.

The news from Japan is no better.

> The most serious problem in the mathematics education in our country is that of dropouts in a great number. Someone said that 30% of pupils drop during the Elementary School, 50% in the Lower Secondary School, and 70% in the Upper Secondary School. (Hirabayashi, 1984, p. 1)

So what has been missing from the reforming movements of the last two decades? Unless we can find at least a partial answer to this question, there is no reason to suppose that future efforts will be any more successful than those of the past. And the effects on our nations' children[1] are too serious to be ignored. Simply to increase the pressures on them will only make things worse.

> these students are unable to understand what school math is all about (with very real cause); they are now wrongly judged as "witholding . . . efforts" and are demanded to do far more of the same work. This cannot but throw great numbers, already with great math anxiety, into severe crisis. (Whitney, 1985, p. 123)

So where are we still going wrong?

I don't think that there is a single answer to the problem, nor that any one person knows all the answers. As my own contribution to progress, I have two answers to offer. The present book centres on one of these; and at the end of this introductory Chapter, I shall indicate in what directions we should look for an answer to the other.

---

[1]Elsewhere (Skemp, 1979a. Chapter 16), I distinguish between those who enter a learner–teacher relationship of their own free choice, and those who attend school because they must. The former relationship I call that of *student* and *mentor,* and the latter that of *pupil* and *schooler.* These are used as technical terms, and I am aware that this is not the everyday usage, especially in the United States; but I see the distinction as important, and ask readers to bear with me. For the most part I avoid the use of the word "student" altogether, except in passages quoted from other authors; and talk about children, learners, and (where it fits the meaning) pupils.

I came to my first answer by a long and roundabout path. Here, a few paragraphs of an autobiographical nature are necessary.

## A MENTAL JOURNEY AROUND THE WORLD

The English poet John Betjeman once said that the best way to appreciate London is to take a journey round the world, starting in London and finishing in London. My present journey has taken more than 30 years. It began in, and has returned to, the mathematics classroom. But, meanwhile, it has taken me into the areas of developmental psychology, motivation, human emotions, cybernetics, evolution, and human intelligence. Eventually it led me to reformulate my own conception of human intelligence, at which point I found myself, unexpectedly, back at mathematics again.

This journey was largely a quest for solutions of two problems, one professional and one theoretical. The professional problems arose out of my job as a teacher, trying to teach mathematics and physics to children from 11 upwards. Over a period of five years at this, I became increasingly aware that I wasn't being as successful as I wished. Some pupils did well, but others seemed to have a blockage for maths. It was not lack of intelligence or hard work, on their side or mine. So we had a problem: a problem not confined to myself as a teacher, nor to these particular children as pupils. This was in the late 1940s: since then, awareness of this problem has become widespread.

An outcome of this was that I became increasingly interested in psychology, went back to university, and took a psychology degree. Problems of learning and teaching are psychological problems, so it was reasonable to expect that by studying psychology I would find answers to my professional problems as a teacher.

Unfortunately I didn't. Learning theory at that time was dominated by behaviourism; theories of intelligence were dominated by psychometrics. Neither of these theories (or groups of theories) was of any help in solving my professional problems, as a teacher. This I had come to realise long before the end of my degree course, during which I was still teaching part-time to support myself.

So now I also had a theoretical problem, namely that of finding a theory appropriate to the learning of mathematics. It turned out to be a do-it-yourself job, which is partly why it took so long. The other reason is conveyed by the well known joke about the person who asked the way to (let us call it) Exville, and was told "If I wanted to get to Exville, I wouldn't start from here." I too was starting from the wrong beginnings. Behaviourist models are helpful in understanding those forms of learning that we have in common with the laboratory rat and pigeon; and it has to be admitted that for too many children, the word "mathematics" has become a conditioned anxiety stimulus. But the learning of mathematics with understanding exemplifies the kind of learning in which humans most differ from the lower animals: so for this we need a different kind of theoretical model. And psychometric models attempt to tell us how much intelligence a person has. They do not tell us what it is they have this amount *of;* nor do they relate to the process of learning. The use of a noun here tends to mislead, unless it is expanded. It is helpful to compare this with our use of the word "memory." When we say that someone has a good memory, we mean that this person is well able to take in information, organise it, store it, and retrieve from his large memory store just what he needs at any particular time. We are talking about a cluster of mental abilities, which are very useful. So I am suggesting that we need to think about intelligence in the same sort of way: as a cluster of mental abilities which, collectively, are very useful.

If we continue along this line of thinking, the next question that arises is, what are these abilities that collectively comprise the *functioning* of human intelligence? If we can answer some of these questions, we shall be on the way to relating intelligence and learning.

## MATHEMATICS AND HUMAN INTELLIGENCE

All the time I was working on the psychology of learning maths, I was also working on the psychology of intelligent learning. Initially I did not realise this. But gradually I came to regard maths as a particularly clear and concentrated example of the activity of human intelligence, and to feel a need to generalise my thinking about maths learning into a theory for intelligent learning that would be applicable to all subjects: and for teaching that would help this to take place. For it became ever more clear that mathematics was not the only subject that was badly taught and ill understood: it just showed up more clearly in mathematics.

The desire became intensified in 1973, when I moved from the University of Manchester to that of Warwick, and from a Psychology to an Education department. Over the next five years I continued to work on this, and the outcome has been nothing less than a new model for intelligence itself (Skemp, 1979a).

This is an ambitious undertaking. But the earlier models, based on IQ and its measurement, have been with us now for about 70 years; and although they may have developed much expertise in the measurement of intelligence, they tell us little or nothing about how it functions, why it is a good thing to have, and how to make the best use of whatever intelligence we have. Until we turn our thoughts from measurement to function, the most important questions about intelligence will remain not only unanswered, but barely even asked.

Like the traveller returning to London, I now see mathematics in a new perspective. First, I see it as a particularly powerful and concentrated example of the *functioning* of human intelligence. And second, I see it as one of the most powerful and adaptable mental tools which the intelligence of man has made for its own use, collectively over the centuries. There is a close analogy between this and the use of our hands to make physical tools. We can do quite a lot with our bare hands, directly on the physical world. But we also use our hands to make a variety of tools—screwdrivers, cranes, lathes—and these greatly amplify the abilities of our hands. This is an indirect activity, but longterm it is exceedingly powerful. Likewise, mathematics is a way of using our minds that greatly increases the power of our thinking. Hence, it is important in today's world of rapidly advancing science, high technology, and commerce.

If this view is correct, then it is predictable that children will not succeed in learning maths unless they are taught in ways that enable them to bring their intelligence, rather than rote learning, into use for their learning of mathematics. This was not, and still is not, likely to happen as long as for the majority of educational psychologists, and those who listened to them, intelligence is so closely linked with IQ that the two are almost synonymous. We thus return to the question: What are the activities that collectively make up the functioning of intelligence? We need at least some of the answers to this question before we can begin to devise learning situations that evoke these activities (i.e., which evoke intelligent learning). The remaining chapters of this book are given to consideration of this question: the functioning of intelligence, as it relates to the learning of mathematics.

## THE SCHOOL SITUATION

Before leaving this chapter, however, I must keep my promise to mention the other direction in which I believe change is needed. During an eight-year project for the development of methods and materials for putting the present theory to use in the classroom, I became more and more aware of the crucial importance of the school environment. In some schools, these methods took root and flourished. But in others, they never took root at all; or if they did, showed no growth and gradually withered away. My gratitude to those schools in the first category is great because without them I would have become very discouraged. To visit these schools, and talk and work with their teachers and children, had been a source of great intellectual and personal pleasure and reward. But I am saddened by the other schools because it is not only for mathematics that their climate is unhelpful. In these schools there is a hidden curriculum, which favours obedience and teacher-dependence rather than cooperation and the development of children's own powers of thought. Space, and the title of this book, do not allow me to develop this theme further (see, for example, Holt, 1969). The two contrasting ways of thinking are well exemplified by two further quotations.

In *A Nation at Risk* (National Commission on Excellence in Education, 1983) we read "Students in high schools should be assigned far more homework than is now the case." And later, "To Students: You forfeit your chance for life at its fullest when you withold your best efforts in learning . . . ." (Whitney, 1985, p. 123)

In contrast:

Teachers face a dilemma when they try to move children to do school work that is not intrinsically interesting. Children must be induced to undertake the work either by promise of reward or threat of punishment, and in neither case do they focus on the material to be learnt. In this sense the work is construed as a bad thing, an obstacle blocking the way to reward of a reason for punishment. Kurt Lewin explores this dilemma in "The Psychological Situation of Reward and Punishment" (*A Dynamic Theory of Personality: Selected Papers of Kurt Lewin*, McGraw-Hill, 1935). The ideas of Piaget and Lewin have led me to state the central problem of education thus: How can we instruct while respecting the self-constructive character of mind? (Lawler, 1982, p. 138.)

## OVERVIEW

The rest of this book is about human intelligence and the learning of mathematics. Part A, Chapters 2 to 7, are as they appeared in the original edition published by Penguin in 1971. Since these chapters first appeared, there has been an increasing amount of valuable work in this field, much of it inspired by the pioneering work of Piaget. If I were starting now from the beginning, there would of course be many references to this work. The result would be quite a different book, more in the nature of a survey; and there are already books in print which do this different job well. But because there is nothing in the original chapters about which I now think differently, it has seemed better not to risk changing what is still being well received in seven languages, but to add a sequel. This forms Part B.

The order in which the chapters now appear is one good order in which to read them. However, for those entirely new to the subject, Chapter 12 provides a good introduction. Since it first appeared in the journal *Mathematics Teaching,* this has been read by more people than anything else I have written, and it fits well with the intuitions of many. The full theoretical underpinning for these ideas will be understood later, when returning to this chapter in its numerical sequence. Another good order would be to read chapter 8 first, and then the earlier chapters. This plan will allow the reader to see where I was going somewhat better than, at the time, I did myself.

# 2

# The Formation
# of Mathematical
# Concepts

In this chapter we shall examine what we mean by concepts, and how we form, use and communicate these. Then, in Chapter 3, we shall consider how concepts fit together to form conceptual structures, called schemas, and examine some of the results which follow from the organization of our knowledge into these structures.

## ABSTRACTING AND CLASSIFYING

Though the term 'concept' is widely used, it is not easy to define. Nor, for reasons which will appear later, is a direct definition the best way to convey its meaning. So I shall approach it from several directions, and with a variety of examples. Since mathematical concepts are among the most abstract, we shall reach these last.

First, two pre-verbal examples. A baby boy aged twelve months, having finished sucking his bottle, crawled across the floor of the living room to where two empty wine bottles were standing and stood his own empty feeding bottle neatly alongside them. A two-year-old boy, seeing a baby on the floor, reacted to it as he usually did to dogs, patting it on the head and stroking its back. (He had seen plenty of dogs, but had never before seen another baby crawling.)

In both these cases the behaviour of the children concerned implies: one, some kind of classification of their previous experience; two, the fitting of their present experience into one of these classes.

We all behave like this all the time; it is thus that we bring to bear our past experiences on the present situation. The activity is so continuous and automatic that it requires some slightly unexpected outcome thereof, such as the above, to call it to our attention.

At a lower level we classify every time we recognize an object as one which we have seen before. On no two occasions are the incoming sense-data likely to be exactly the same, since we see objects at different distances and angles, and also in varying lights. From these varying inputs we abstract certain *in*variant properties, and these properties persist in memory longer than the memory of any particular presentation of the object. In the diagram, $C_1, C_2$ . . . represent suc-

cessive past experiences of the same object, say, a particular chair. From these we abstract certain common properties, represented in the diagram by $C$. Once this abstraction is formed, any further experience, $C_n$, evokes $C$, and the chair is *recognized:* that is, the new experience is classified with $C_1, C_2$, etc.; $C_n$ and $C$ are now experienced together; and from their combination we experience both the *similarity* ($C$) of $C_n$ to our previous experiences of seeing this chair and also the particular distance, angle, etc., on this occasion ($C_n$).

We progress rapidly to further abstractions. From particular chairs, $C, C', C''$, we abstract further invariant properties, by which we recognize $Ch$ (a new object seen for the first time, say, in a shop window) as a member of this class. It is the second-order abstraction (from the set of abstractions $C, C', \ldots$) to which we give the name 'chair.' The invariant properties which characterize it are already

becoming more functional and less perceptual—that is, less attached to the physical properties of a chair. One I saw recently was of basket-work, egg-shaped and hung from a single rope. It bore little or no physical resemblance to any chair which I had ever seen—but I recognized it at once as a chair, and a most desirable one too!

From the abstraction *chair*, together with other abstractions such as *table*, *carpet*, *bureau*, a further abstraction, *furniture*, can be made, and so on. These classifications are by no means fixed. Particularly by the young, chairs are also classified as things to stand on, gymnastic apparatus and framework of a play house. Tables are sometimes used as chairs. This flexibility of classification, according to the needs of the moment, is clearly an aid to adaptability.

Naming an object classifies it. This can be an advantage or a disadvantage. A very important kind of classification is by function, and once an object is thus classified, we know how to behave in relation to it. 'Whatever is this?' 'It's a gadget for pulling off Wellington boots.' But once it is classified in a particular way, we are less open to other classifications. Most of us classify cars as vehicles, time-savers and perhaps status symbols, and use them in accordance with these functions. Fewer also see them as potentially lethal objects, and our behaviour therefore takes less account than it should of this classification.

It may be useful to bring together some of the terms used so far. *Abstracting* is an activity by which we become aware of similarities (in the everyday sense) among our experiences. *Classifying* means collecting together our experiences on the basis of these similarities. An *abstraction* is some kind of lasting mental change, the result of abstracting, which enables us to recognize new experiences as having the similarities of an already formed class. Briefly, it is something learnt which enables us to classify; it is the defining property of a class. To distinguish between abstracting as an activity and an abstraction as its end-product, we shall hereafter call the latter a *concept*.

A concept therefore requires for its formation a number of experiences which have something in common. Once the concept is formed, we may (retrospectively and prospectively) talk about *examples* of the concept. Everyday concepts come from everyday experience, and the examples which lead to their formation occur randomly, spaced in time. More frequently encountered objects are, in general, conceptualized more rapidly; but in many other factors are at work which make this statement an oversimplification. One of these is *contrast*.

In the diagram on the right the single $X$ stands out perceptually from the five variously shaped $O$s. Objects which thus stand out from their sur-  roundings are more likely to be remembered and their similarities are more likely to be abstracted across intervals of space and time.

The diagram also illustrates the function of non-examples in determining a class. The $X$, by its difference from all the other shapes, makes the similarity between them more noticeable. The essential characteristics of a *chair* are clarified by pointing to, say, a stool, a settee, a bed and a garden seat, and saying

'These are not chairs.' This is specially useful in fixing the borderline of a class—we use objects which might be examples, but aren't.

## NAMING

We have just used *naming* again. Language is, in humans, so closely linked with concepts and concept-formation that we cannot for long keep naming out of our discussion. Indeed, many people find it difficult to separate a concept from its name, as is shown by the following charming illustration provided by Vygotsky (1962). Children were told that in a game a dog would be called 'cow.' The following is a typical sequence of questions and answers. 'Does a cow have horns?' 'Yes.' 'But don't you remember that the cow is really a dog? Come now, does a dog have horns?' 'Sure, if it is a cow, if it's called cow, it has horns. That kind of dog has got to have little horns.' Vygotsky also quotes a story about a peasant who, after listening to two students of astronomy talking about the stars, said that he could understand that with the help of instruments people could measure the distance from the earth to the stars and find their positions and motion. What puzzled him was how in the devil they found out the *names* of the stars!

The distinction between a concept and its name is an essential one for our present discussion. A concept is an idea; the name of a concept is a sound, or a mark on paper, associated with it. This association can be formed after the concept has been formed ('What is this called?') or in the process of forming it. If the same name is heard or seen each time, an example of a concept is encountered, by the time a concept is formed, the name has become so closely associated with it that it is not only by children that it is mistaken for the concept itself. In particular, numbers (which are mathematical concepts) and numerals (the names we use for numbers) are widely confused. This point is discussed further in Chapter 4.

Being associated with a concept, the use of a name in connection with an object helps us to classify it, that is, to recognize it as belonging to an existing class. 'What's this?' 'A new kind of bottle opener which works by compressed air.' Now we have classified it, which we were unable to do by its perceptual properties alone; so we know what to do with it. This classification was done by bringing the concept of a bottle opener to consciousness at the same time as the new experience.

Naming can also play a useful, sometimes an essential, part in the formation of new concepts. Hearing the same name in connection with different experiences predisposes us to collect them together in our minds and also increases our chance of abstracting their intrinsic similarities (as distinct from the extrinsic one of being called by the same name). Experiment has also shown that associating different names with classes which are only slightly different in their characteristics helps to classify later examples correctly, even if the later examples are not named. The names help to separate the classes themselves.

## THE COMMUNICATION OF CONCEPTS

We can see that language can be used to speed up the formation of a concept by helping to collect and separate contributory examples and non-examples. Can it be used to short-circuit the process altogether by simply defining a concept verbally? Particularly in mathematics, this is often attempted, so let us examine the idea of a definition, as usual with the help of examples. To begin with, let us choose a simple and well-known concept, say, *red,* and imagine that we are asked the meaning of this word by a man, blind from birth, who has been given sight by a corneal graft. The meaning of a word is the concept associated with that word, so our task is now to enable the person to form the concept *red* (which he does not have when we begin) and associate it with the word 'red.'

There are two ways in which we might do this. Being scientifically inclined, and perhaps interested in colour photography, we could give a definition: 'Red is the colour we experience from light of wavelength in the region of 0·6 microns.' Would he now have the concept red? Of course not. Such a definition would be useless *to him,* though not necessarily for other purposes. Intuitively, in such a case, we would point to various objects and say 'This is a red diary, this is a red tie, this is a red jumper . . .' In this way we would arrange for him to have, close together in time, a collection of experiences from which we hope he will abstract the common property—red. Naming is here used as an auxiliary, in the way already described. The same process of abstraction could take place in silence, but it would probably be slower and the name 'red' would not become attached.

If he now asks a different question, 'What does ''colour'' mean?', we can no longer collect together examples for him by pointing, for the examples we want are *red, blue, green, yellow* . . ., and these are themselves concepts. If, and only if, he already has these concepts in his own mind—their presence in our mind is not enough—then, by collecting together the words for them, we can arrange for him to collect together the concepts themselves, and thus make possible, though not guarantee, the process of abstraction. Naming (or some other symbolization) now becomes an essential factor of the process of abstraction and not just a useful help.

This leads us to an important distinction between two kinds of concept. Those which are derived from our sensory and motor experiences of the outside world, such as *red, motor car, heavy, hot, sweet,* will be called *primary concepts;* those which are abstracted from other concepts will be called *secondary concepts.* If concept $A$ is an example of concept $B$, then we shall say that $B$ is of a higher order than $A$. Clearly, if $A$ is an example of $B$, and $B$ of $C$, then $C$ is also of higher order than both $B$ and $A$. 'Of higher order than' means 'abstracted from' (directly or indirectly). So 'more abstract' means 'more removed from experience of the outside world,' which fits in with the everyday meaning of the word 'abstract.' This comparison can only be made between concepts in the same hierarchy. Although we might consider that *sonata form* is a more abstract (higher order) concept than *colour*, we cannot properly compare the two.

These related ideas, of order between concepts and a conceptual hierarchy, enable us to see more clearly why, for the person we are thinking of, the definition of red was an inadequate mode of communication: it presupposed concepts such as *colour, light,* which could only be formed if concepts such as *red, blue, green* . . . had already been formed. In general, *concepts of a higher order than those which people already have cannot be communicated to them by a definition* but only by collecting together, for them to experience, suitable examples.

Of what use, then, if any, is a definition?

Two uses can be seen at once. If it were necessary (for example, for a photographic colour filter) to specify exactly within what limits we would still call a colour red, then the above definition would enable us to say where red starts and finishes. And having gone further in the process of abstraction, that is, in the formation of larger classes based on similarities, a definition enables us to retrace our steps. By stating all those (and only those) classes to which our particular concept belongs, we are left with just one possible concept—the one we are defining. In the process we have shown how it relates to the other concepts in its hierarchy. Definitions can thus be seen as a way of adding precision to the boundaries of a concept, once formed, and of stating explicitly its relation to other concepts.

New concepts of a lower order can also be communicated for the first time by this means. For example, if our formerly blind subject asked 'What colour is magenta?' and we could not find a sufficiency of magenta objects to show him, we could say 'It is a colour, between red and blue, rather more blue than red.' Provided that he already had the concepts of blue and red, he could then form at least a beginning of the concept of magenta without ever having seen this colour.

Since most of the new concepts we need in everyday life are of a fairly low order, we usually have available suitable higher-order concepts for the new concepts to be easily communicable by definition, often followed by an example or two, which then serve a different purpose—that of illustration. 'What is a stool?' 'It's a seat without a back for one person' is quite a good definition, but even so a few examples will define the concept in such a way as to exclude hassocks, pouffes and garden swings far more successfully than further elaboration of the definition.

In mathematics, however, not only are the concepts far more abstract than those of everyday life, but the direction of learning is for the most part in the direction of still greater abstraction. The communication of mathematical concepts is therefore much more difficult, on the part of both communicator and receiver. This problem will be taken up again shortly, after certain other general topics have been explored.

## CONCEPTS AS A CULTURAL HERITAGE

Low-order concepts can be formed, and used, without the use of language. The criterion for *having* a concept is not being able to say its name but behaving in a way indicative of classifying new data according to the similarities which go to form this concept. Animals behave in ways from which one may reasonably infer that they form simple concepts. A rat, trained to choose a door coloured mid-grey in preference to a light grey, will if now presented with doors of mid-grey and dark grey go to the dark grey. It processes the data in terms of 'darker than.'

The most obvious discontinuity between human beings and other animals is in the former's use of language. What this implies is less obvious. If we choose a word at random it will almost always be found that the concept which the word names—the meaning of the word—is not a specific object or experience but a class. (Proper nouns are a partial exception.)

Now, there are two ways of evoking a concept, that is, of causing it to start functioning. One is by encountering an example of the concept. The concept then comes into action as our way of classifying this example, and our subjective experience is that of *recognition*. The other is by hearing, reading or otherwise making conscious the name, or other symbol, for the concept. Animals can do the first; only human beings can do the second. And the reason for this lies not in superior vocal apparatus, but *in the ability to isolate concepts from any of the examples which give rise to them*. Only by being detachable from the sensory experiences from which they originated can concepts be collected together as examples from which new concepts of greater abstraction can be formed.

We would expect this detachability to be related to abstracting ability, for the stronger the mental organization based not on direct sense-data but on similiarities between them, the greater we would expect its ability to function as an independent entity. This view is supported by evidence from several sources. Children of very low intelligence do not learn to talk, in spite of adequate vocal apparatus. Chimpanzees, the closest of our surviving ancestors, can learn to sit at a table and drink from a cup, but not to talk. Human beings are the most intelligent and the most adaptable of all species. They are also the only species who can talk.

Our ability to make concepts independent of the experiences which gave rise to them and to manipulate them by the use of language is the very core of human superiority over other species. This is the first step towards the realization of the potential which this greater intelligence gives. Intelligence makes speech possible, but speech (which has to be learnt) is essential for the formation and use of the higher-order concepts which, collectively, form our scientific and cultural heritage.

A concept is a way of processing data which enables the user to bring past experience usefully to bear on the present situation. Without language each individual has to form his own concept direct from the environment. Without language, these primary concepts cannot be brought together to form concepts of higher order. By language, however, the first process can be speeded up and the second is made possible. Moreover, the concepts of the past, painstakingly abstracted and slowly accumulated by successive generations, become available to help each new individual form his own conceptual system.

The actual construction of a conceptual system is something which individuals have to do for themselves. But the process can be enormously speeded up if, so to speak, the materials are to hand. It is like the difference between building a boat from a kit of wood already sawn to shape and having to start by walking to the forest, felling the trees, dragging them home, making planks— having first mined some iron ore and smelted it to make an axe and a saw!

What is more, the work of geniuses can be made available to everyone. Concepts like that of gravitation, the result of years of study by one of the greatest intelligences the world has known, become available to all scientists who follow. The first person to form a new concept of this order has to abstract it relatively unaided. Thereafter, language can be used to direct the thoughts of those who follow so that they can make the same discovery in less time and with less intelligence. Yet even Newton (1642–1727) was by no means altogether unaided. He said, with modesty, 'If I have seen a littler farther than others, it is because I have stood on the shoulders of giants.' The conceptual structures of earlier mathematicians and scientists were available to him.

In this context, the generalized idea of *noise* is useful. By this is meant data which is irrelevant to a particular communication, so that what is noise in one context may not be so in another. (For example, if we are listening to music when the telephone rings, the sound of the bell conveys *information* that someone is calling us, but is *noise* relative to the music.) The greater the noise, the harder it is to form the concept. Before reading on, please put your hand over the diagrams which are on the right-hand side on the next page. Try to form the concept from the high-noise examples and non-examples. Now remove your hand and try to form the concept from the low-noise examples of the same concept.

From the right-hand examples it is much easier to see that the concept is *having intersecting lines*. The extra noise in the left-hand examples comes partly from the additional lines, but largely from the fact that each looks like something.

An attribute of high intelligence is the ability fo form concepts under conditions of great noise. But once we have a concept, we can see examples of it where previously we could not.

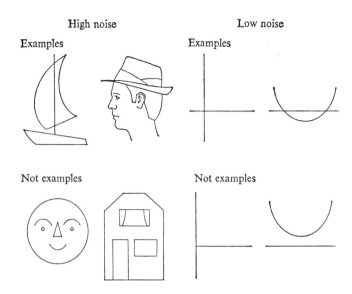

## THE POWER OF CONCEPTUAL THINKING

Conceptual thinking confers on users great power to adapt their behaviour to the environment, and to shape their environment to suit their own requirements. This results partly from the detachment of the concepts from both present sense-data and behaviour, and their manipulation independently of these. We take this so much for granted that we hardly realize the enormous advantage of *not* having to do something in order to discover whether it is the best thing to do! But, of course, all major activities, from setting up in business to building an aircraft, are put together in thought before they are constructed in fact.

The power of concepts also comes from their ability to combine and relate many different experiences and classes of experience. The more abstract the concepts, the greater their power to do this. The person who says 'Don't worry me with theory—just give me the facts' is speaking foolishly. A set of facts can be used only in the circumstances to which they belong, whereas an appropriate theory enables us to explain, predict and control a great number of particular events in the classes to which it relates.

A further contribution to the power of conceptual thinking is related to the shortness of our span of attention. Our short-term memory can only store a limited number of words or other symbols. Clearly the higher the order of the concepts which these symbols represent, the greater the stored experience they bring to bear. Mathematics is the most abstract, and so the most powerful, of all theoretical systems. It is therefore potentially the most useful; scientists in particular, but also economists and navigators, businessmen and communications engineers, find it an indispensable 'tool' (data-processing system) for their work. Its usefulness is, however, only potential, and many who work wearily at trying to learn it throughout their schooldays derive little benefit, and no enjoyment. This is almost certainly because they are not really learning mathematics at all. The latter is an interesting and enjoyable process, though many will find this hard to believe. What is inflicted on all too many children and older students is the manipulation of symbols with little or no meaning attached, according to a number of rote-memorized rules. This is not only boring (because meaningless); it is very much harder, because unconnected rules are much harder to remember than an integrated conceptual structure. The latter point will be taken up in the next chapter. Here we shall concentrate on the communication of mathematical concepts.

## THE LEARNING OF MATHEMATICAL CONCEPTS

Much of our everyday knowledge is learnt directly from our environment, and the concepts involved are not very abstract. The particular problem (but also the power) of mathematics lies in its great abstractness and generality, achieved by successive generations of particularly intelligent individuals each of whom has been abstracting from, or generalizing, concepts of earlier generations. The present-day learner has to process not raw data but the data-processing systems of existing mathematics. This is not only an immeasurable advantage, in that an able student can acquire in years ideas which took centuries of past effort to develop; it also exposes the learner to a particular hazard. Mathematics cannot be learnt directly from the everyday environment, but only indirectly from other mathematicians, in conjunction with one's own reflective intelligence. At best, this makes one largely dependent on teachers (including all who write mathematical textbooks); at worst, it exposes one to the possibility of acquiring a lifelong fear and dislike of mathematics.

Though the first principles of the learning of mathematics are straightforward, it is the communicator of mathematical ideas, and not the recipient, who most needs to know them. And though they are simple enough in themselves, their mathematical applications involve much hard thinking. The first of these principles was stated earlier in the chapter:

*(1) Concepts of a higher order than those which people already have cannot be communicated to them by a definition, but only by arranging for them to encounter a suitable collection of examples.*

The second follows directly from it:

*(2) Since in mathematics these examples are almost invariably other concepts, it must first be ensured that these are already formed in the mind of the learner.*

The first of these principles is broken by the vast majority of textbooks, past and present. Nearly everywhere we see new topics introduced not by examples but by definitions, definitions of the most admirable brevity and exactitude for the teacher (who already has the concepts to which they refer) but unintelligible to the student. For reasons which will be apparent, examples cannot be quoted here, but readers are invited to verify this statement for themselves. It is also a useful exercise to look at some definitions of ideas new to oneself in books about mathematics beyond the stage which one has reached. This enables one to experience at first hand the bafflement of the younger learner.

Good teachers intuitively help out a definition with examples. To choose a suitable collection is, however, harder than it sounds. The examples must have in common the properties which form the concept but no others. To put it differently, they must be alike in the ways which are to be abstracted, and otherwise different enough for the properties irrelevant to this particular concept to cancel out or, more accurately, fail to summate. Remembering that these irrelevant properties may be regarded as noise, we may say that some noise is necessary to concept formation. In the earlier stages, low noise—clear embodiment of the concept, with little distracting detail—is desirable; but as the concept becomes more strongly established, increasing noise teaches the recipient to abstract the conceptual properties from more difficult examples and so reduces dependence on the teacher.

Composing a suitable collection thus requires both inventiveness and a very clear awareness of the concept to be communicated. Now, it is possible to have, and use, a concept at an intuitive level without being consciously aware of it. This applies particularly to some of the most basic and frequently used ideas: partly because the more automatic any activity, the less we think about it; partly because the most fundamental ideas of mathematics are acquired at an early age, when we have not the ability to analyse them; and partly because some of these fundamental ideas are also among the most subtle. But it is easy to slip up even when these factors do not apply.

Some children were learning the theorem of Pythagoras (c. sixth century BC). They had copied a right-angled triangle from the blackboard—figure a—and were told to make a square on each side. This they did easily enough for the two shorter sides—figure b; but they were nearly all in difficulty when they tried to draw the square on the hypotenuse. Many of them drew something like figure c. From this, I inferred that the squares from which they had formed their concepts had all been 'square' to the paper and had included no obliquely placed examples. All too easily done!

The second of the two principles, that the necessary lower-order concepts must be present before the next stage of abstraction is possible, seems even more straightforward. To put this into effect, however, means that before we try to communicate a new concept, we have to find out what are its contributory concepts; and for each of these, we have to find out *its* contributory concepts,

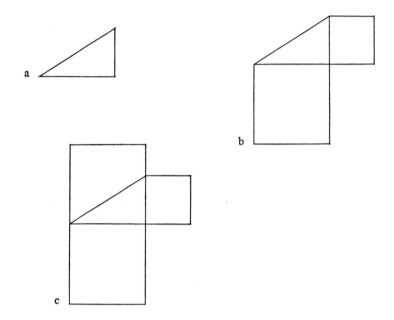

and so on, until we reach either primary concepts or experience which we can assume. When this has been done, a suitable plan can then be made which will present to the learner a possible, and not an impossible, task.

This conceptual analysis involves much more work than just giving a definition. If done, it leads to some surprising results. Ideas which not long ago were first taught in university courses are now seem to be so fundamental that they are being introduced in the primary school: for example, sets, one-to-one correspondence. Other topics still regarded as elementary are found on analysis to involve ideas which even those teaching the topic have for the most part never heard of. In this category I include the manipulation of fractional numbers.

There are two other consequences of the second principle. The first is that in the building up of the structure of successive abstractions, if a particular level is imperfectly understood, everything from then on is in peril. This dependency is probably greater in mathematics than in any other subject. One can understand the geography of Africa even if one has missed that of Europe; one can understand the history of the nineteenth century even if one has missed that of the eighteenth; in physics one can understand 'heat and light' even if one has missed 'sound.' But to understand algebra without ever having really understood arithmetic is an impossibility, for much of the algebra we learn at school is generalized arithmetic. Since many pupils learn to do the manipulations of arithmetic with a very imperfect understanding of the underlying principles, it is small wonder that mathematics remain a closed book to them. Even those who get off to a good start may, through absence, inattention, failure to keep up with the pace of the class or other reasons, fail to form the concepts of some particular stage. In that case, all subsequent concepts dependent on these may never be understood, and pupils become steadily more out of their depth. In the latter case, however, the situation may not be so irremediable, if the learning situation is one which makes back-tracking possible: for example, if the text in use provides a genuine explanation and is not just a collection of exercises. Success will then depend partly on the confidence of the learners in their own powers of comprehension.

The other consequences (of the second principle) is that the contributory concepts needed for each new stage of abstraction must be *available*. It is not sufficient for them to have been learnt at some time in the past; they must be accessible when needed. This is partly a matter, again, of having facilities available for back-tracking. Appropriate revision, planned by a teacher, will be specially useful for beginners, but more advanced students should be taking a more active part in the direction of their own studies, and, for these, returning to take another look at earlier work will be more effective if it is directed by a felt need rather than by an outside instruction. To put it differently, an answer has more meaning to someone who has first asked a question.

## LEARNING AND TEACHING

In learning mathematics, although we have to create all the concepts anew in our own minds, we are only able to do this by using the concepts arrived at by past mathematicians. There is too much for even a genius to do in a lifetime.

This makes the learning of mathematics, especially in its early stages and for the average student, very dependent on good teaching. Now, to know mathematics is one thing and to be able to teach it—to communicate it to those at a lower conceptual level—is quite another; and I believe that it is the latter which is most lacking at the moment. As a result, many people acquire at school a lifelong dislike, even fear, of mathematics.

It is good that widespread efforts have been and are still being made to remedy this, for example, by the introduction of new syllabi, more attractive presentation, television series and other means. But the small success of these efforts, after twenty years or more, supports the view already put forward in the introduction, namely; that these efforts will be of little value until they are combined with greater awareness of the mental processes involved in the learning of mathematics.

# The Idea of a Schema

Though in the previous chapter our attention was centred on the formation of single concepts, each of these by its very nature is embedded in a structure of other concepts. Each (except primary concepts) is derived from other concepts and contributes to the formation of yet others, so it is part of a hierarchy. But at each level alternative classifications are possible, leading to different hierarchies. A car can be classed as a vehicle (with buses, trains, aircraft), as a status symbol (with a title, a good address, a mink coat), as a source of inland revenue (with tobacco, drink, and dog licenses), as an export (with gramophone records, Scotch whisky, Harris tweed), etc. What is more, the class concepts on which we have been concentrating so far are by no means the only kind. Given a collection not of single objects but of *pairs* of objects we may become aware of something in common between the pairs. For example:

puppy, dog; kitten, cat; chicken, hen.

Here we see that each of these pairs can be connected by the idea '. . . is a young . . .' Another example:

Bristol, England; Hull, England; Rotterdam, Holland.

In this, each pair can be connected by the idea '. . . is a port of . . .' These two connecting ideas are themselves examples of a new idea called a *relation*. A mathematical relation may be seen in the following collection of pairs.

$$(6, 5), \quad (2, 1), \quad (9, 8), \quad (32, 31) \ldots$$

We can call this relation 'is one more than' or 'is the successor of.' Another mathematical example:

$$(\tfrac{1}{2}, \tfrac{2}{4}), \qquad (\tfrac{1}{3}, \tfrac{2}{6}), \qquad (\tfrac{1}{4}, \tfrac{2}{8}) \ldots$$

This relation is called 'is equivalent to'. The fractions in each pair, though not identical, represent the same number. Notice (1) that in mathematics it is usual to enclose the pairs in a given relation in parentheses, as above; (2) that the order within the pairs usually matters. These:

$$(5, 6), \qquad (1, 2), \qquad (8, 9), \qquad (31, 32)$$

are in a different relation to these:

$$(6, 5), \qquad (2, 1), \qquad (9, 8), \qquad (32, 31)$$

We can even start to classify these relations. Those mathematical relations given as examples in the last paragraph were chosen to exemplify two particular kinds: order relations and equivalence relations. Other order relations are: is greater than, is the ancestor of, happened after. Other equivalence relations are: is the same size as, is the sibling of, is the same colour as. Both order relations and equivalence relations have important general properties. So we have not only a hierarchical structure of class concepts but another structure of individual relations, and classes of relations, which forms cross-linkages within the first structure.

Another source of cross-linkages arises from our ability to 'turn one idea into another' by doing something to it.

| | | | |
|---|---|---|---|
| *Example:* | good→bad | hot→cold | high→low |
| *Another example:* | good→best | bad→worst | high→highest |

This 'something which we can do to an idea' is called a *transformation,* or more generally a *function.* There are many different kinds of transformation, and, what is more, we can on occasion combine two particular transformations to get another transformation (just as one can combine two numbers to get another). For example, by combining the two transformations above we get

$$\text{good} \rightarrow \text{worst}, \quad \text{hot} \rightarrow \text{coldest}, \qquad \text{etc.}$$

So transformations are both connected among themselves and are also another source of connections between the ideas to which they can be applied.

The foregoing offers a brief, and perhaps rather concentrated, glimpse of the richness and variety of the ways in which concepts can be interrelated, and of the resulting structures. The study of the structures themselves is an important part of mathematics, and the study of the ways in which they are built up and function is at the very core of the psychology of learning mathematics.

Now, when a number of suitable components are suitably connected, the resulting combination may have properties which it would have been difficult to predict from a knowledge of the properties of the individual components. How many of us could have predicted from knowledge of the separate properties of transitors, condensers, resistors and the like that, when these are suitably connected, the result would enable us to hear radio broadcasts?

So it is with concepts and conceptual structures. The new function of the electrical structure described above is marked by a new name—transistor radio. Likewise, a conceptual structure has its own name—*schema*. The term includes not only the complex conceptual structures of mathematics but also relatively simple structures which coordinate sensori-motor activity. Here we shall be concerned mainly with abstract conceptual schemas. The previous chapter has shown that these concepts have their origins in sensory experience of, and motor activity towards, the outside world. But they soon become detachable from their origins, and their further development takes place by interaction with other mathematicians and with each other.

Among the new functions which a schema has, beyond the separate properties of its individual concepts, are the following: it integrates existing knowledge, it acts as a tool for future learning and it makes possible understanding.

## THE INTEGRATIVE FUNCTION OF A SCHEMA

When we recognize something as an example of a concept we become aware of it at two levels: as itself and as a member of this class. Thus, when we see some particular car, we automatically recognize it as a member of the class of private cars. But this class-concept is linked by our mental schemas with a vast number of other concepts, which are available to help us behave adaptively with respect to the many different situations in which a car can form a part. Suppose the car is for sale. Then all our motoring experience is brought to bear, reviews of its performance may be recalled, questions to be asked (m.p.g.?) present themselves. Suppose that the cost is beyond our present bank balance. Then sources of finance (bank loans, hire purchase) come to mind. Suppose, instead, that the car is on the road, but has broken down. Then instruments of help (such as the A.A., nearest garage, telephone boxes) are recalled.

Most of these schemas have probably already been linked with the car concept in the past. But suppose now that we park on a foreshore and find that our wheels have sunk into the soft sand. This presents a problem, to solve which schemas from other fields of experience must be brought to bear, such as the behaviour of tides, ways of making a firm surface on soft sand. The more other schemas we have available, the better our chance of coping with the unexpected. We shall return to this point later in the chapter.

## The Schema as a Tool for Further Learning

Our existing schemas are also indispensable tools for the acquisition of further knowledge. Almost everything we learn depends on knowing something else already. To learn aircraft designing we must know aerodynamics, which depends on prior knowledge of calculus, which requires knowledge of algebra, which depends on arithmetic. To learn advanced physiology requires biochemistry, which needs a knowledge of elementary 'school' chemistry. These, and all higher learning, depend on the basic schemas of reading, writing and speaking (or, exceptionally, communicating in some other way) our native language.

This principle—the dependency of new learning on the availabiity of a suitable schema—is a generalization of the second principle for conceptual learning, stated in Chapter 1 on page 30. In the generalized form, new features become important which were not so noticeable while we were concentrating on the learning of particular concepts, though using hindsight they can be seen to be latent there. As an introduction to these, it will be useful to look at an experiment[1] whose purpose was to try to isolate the factor of a schema in learning, or more precisely, to find out how much difference the presence or absence of a suitable schema made to the amount of new material which could be learnt.

For the purpose of the experiment, an artificial schema was devised, somewhat resembling a Red Indian sign language. On the first day the subjects learnt the meanings of sixteen basic signs, such as:

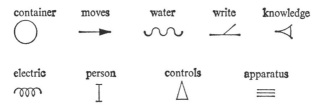

On the second day meanings were assigned to pairs or trios of symbols, such as:

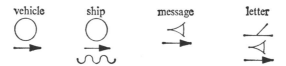

The meanings of these small groups of symbols are related to the meanings of the single symbols, as the reader can verify. On the third and fourth days the groups to be learnt were made progressively larger, the meanings again being related to

---

[1]This is described fully in Skemp (1962).

those of the smaller groups. Here are some examples. (Note that (( )) means plural.)

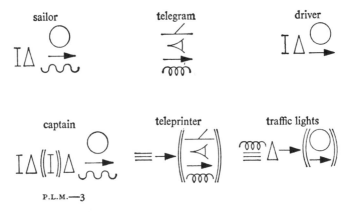

P.L.M.—3

The final task, on the fourth day, was to learn two pages of symbols, each page containing a hundred symbols in ten groups each having from eight to twelve symbols. On one page each group was given a meaning related to the meanings of the smaller groups, as in the examples given. The other page contained groups which were in fact similarly meaningful to a comparison group, but not to these subjects. The comparison group had learnt the same symbols but with different meanings, and these had been built up into a different schema. So in their final task each group had an appropriate schema for one page and an inappropriate schema for the other page. In other words, what was meaningful (in terms of earlier learning) to one group was non-meaningful to the other, and vice versa.

When the results of schematic and 'rote' learning were compared, the differences were striking.

|  | *% recalled (all subjects)* | | |
|---|---|---|---|
|  | *Immediate* | *After one day* | *After four weeks* |
| Schematic | 69 | 69 | 58 |
| Rote | 32 | 23 | 8 |

In this case twice as much was recalled of the schematically learnt as of the rote-learnt material when tested immediately afterwards; and in four weeks the proportion had changed to seven times as much. The schematically learnt material was not only better learnt, but better retained.

Objectively, the two pages of symbols were the same for all the subjects. The only difference was in the mental structures which they had available for the learning task. Clearly, therefore, the schemas which we build up in the course of our early learning of a subject will be crucial to the ease or difficulty with which we can master later topics. When learning schematically—which, in the present context, is to say intelligently—we are not only learning much more efficiently what we are currently engaged in; we are preparing a mental tool for applying the same approach to future learning tasks in that field. Moreover, when subsequently using this tool, we are consolidating the earlier content of the schema. This gives schematic learning a triple advantage over rote memorizing.

There are, however, also certain possible disadvantages to be considered. The first is that, if a task is considered in isolation, schematic learning may take longer. For example, rules for solving a simple equation (see page 86) can be memorized in much less time than it takes to achieve understanding. So if all one wants to learn is how to do a particular job, memorizing a set of rules may be the quickest way. If, however, one wishes to progress, then the number of rules to be learnt becomes steadily more burdensome until eventually the task becomes excessive. A schema, even more than a concept, greatly reduces cognitive strain. Moreover, in most mathematical schemas, all the main contributory ideas are of very general application in mathematics. Time spent in acquiring them is not only of psychological value (meaning that present and future learning is easier and more lasting) but of mathematical value (meaning that the ideas are also of great importance mathematically). In the present context, good psychology is good mathematics.

The second disadvantage is more far-reaching. Since new experience which fits into an existing schema is so much better remembered, a schema has a highly selective effect on our experience. What does *not* fit into it is largely not learnt at all, and what is learnt temporarily is soon forgotten. So, not only are unsuitable schemas a major handicap to our future learning, but even schemas which have been of real value may cease to become so if new experience is encountered, new ideas need to be acquired, which cannot be fitted in to an existing schema. A schema can be as powerful a hindrance as help if it happens to be an unsuitable one.

This brings us to a consideration of adaptability at a new level. So far a schema has been seen as a major instrument of adaptability, being the most effective organization of existing knowledge both for solving new problems and for acquiring new knowledge (and thereby for solving still more new problems in the future). But its very strength now appears as its potential downfall, in that a strong tendency emerges towards the self-perpetuation of existing schemas. If situations are then encountered for which they are not adequate, this stability of the schemas becomes an obstacle to adaptability. What is then necessary is a change of structure in the schemas: they themselves must adapt. Instead of a stable, growing schema by means of which the individual organizes past experience and *assimilates* new data, *reconstruction* is required before the new situation can be understood. This may be difficult, and if it fails, the new experience can no longer be successfully interpreted and adaptive behaviour breaks down— the individual cannot cope.

An everyday example will illustrate the idea, after which some mathematical examples will be given. Early in life, a child learns to distinguish between compatriots and foreigners. His schema of a foreigner is that of a person who comes from abroad, who speaks English with a different accent from his own, perhaps only with difficulty, whose own language is novel and usually incomprehensible, whose mode of dress and personal appearance are slightly or very different. New individual foreigners and new classes—people from countries he had never heard of—are easily assimilated to this concept, which leads to expansion of his schema. But suppose now that he takes a holiday abroad with his parents and discovers that he himself is described as a foreigner. To him, this is incomprehensible. The local inhabitants are the foreigners; he is British! Before he can comprehend this new experience—assimilate it to his schema— the schema itself has to be restructured. His idea of foreigners has to become that of people in a country which is not their own. Not only does this new concept enable him to understand the new experience and so to behave appropriately; it includes the earlier concept as a special case. This is the best kind of reconstruction.

A schema is of such value to an individual that the resistance to changing it can be great, and circumstances or individuals imposing pressure to change may be experienced as threats—and responded to accordingly. Even if it is less than a threat, reconstruction can be difficult, whereas assimilation of a new experience to an existing schema gives a feeling of mastery and is usually enjoyed.

One of the most basic mathematical schemas which we learn is that of the natural number system—the set of counting numbers together with the operations of addition and multiplication. Having learnt to count to ten, a child rapidly progresses to twenty, and is eager to continue the process. Adding single-figure numbers, with the help of concrete materials, is soon learnt. Extending this to the addition of two-figure numbers requires, first, an understanding of our system of numeration based on place value, but once this has been mastered, addition of three-, four-, five-figure numbers is again a straightforward extension. Multiplication is like repeated addition, long multiplication extends simple multiplication. Throughout, the process is one of expansion.

It is another matter when fractional numbers are encountered. These constitute a new number system, not an extension of one which is known already. The system of numeration is different in itself and has new characteristics: for example, an infinite number of different fractions can be used to represent the same number. Multiplication can no longer be understood in terms of repeated addition. Before fractional numbers can be understood, a major reconstruction of the number schema is required. Some people, indeed, go through life without ever really understanding fractional numbers, and small blame to them. Their teacher probably never understood them either, and the difficulty of this particular reconstruction is such that it would require a child of genius level to achieve it unaided at the age when this task is encountered.

The history of mathematics contains some interesting examples showing the difficulty of reconstruction presented by a new number system. When Pythagoras discovered that the length of the hypotenuse of a right-angled triangle could not always be expressed as a rational number, he swore the members of his school to secrecy about this threat to their existing ways of thinking. In his well-known history of mathematics, Bell (1937) says: 'When negative numbers first appeared in experience, as in debits instead of credits, they, as *numbers,* were held in the same abhorrence as ''unnatural'' monstrosities as were later the ''imaginary'' numbers $\sqrt{-1}$, $\sqrt{-2}$, etc.' The Hindu-Arabic system of numerals for the natural numbers also met with great resistance when it was first introduced into Europe in the thirteenth century, and in some places its use was even made illegal. Unspeakable, unnatural, illegal—these are the ways in which the ordinary working tools of present-day mathematics were all characterized by some of the mathematicians who first encountered them. But now that we know the importance of our personal schemas to us, we can begin to understand the defensive nature of these reactions to any new ideas which threaten to overthrow them.

## UNDERSTANDING[2]

We are not in a position to say what we mean by understanding. *To understand something means to assimilate it into an appropriate schema.* This explains the subjective nature of understanding and also makes clear that this is not usually an all-or-nothing state. We may achieve a subjective feeling of understanding by assimilation to an inappropriate schema—the Greeks 'understood' thunderstorms by assimilating these noisy affairs to the schema of a large and powerful being, Zeus, getting angry and throwing things. In this case, an appropriate schema involves the idea of an electric spark, so it was not until the eighteenth century that any real understanding of thunderstorms was possible. The first and major step was taken by Benjamin Franklin, who assimilated concepts about thunderstorms to those about electrical discharges. Fuller understanding, however, involves knowledge of ionization processes in the atmosphere—assimilation to a more extensive schema. What happens in a case like this is that the basic schema becomes enlarged and to the original points of assimilation—noise to noise, lightning flash to electric spark—more are added. Better internal organization of a schema may also improve understanding, and clearly there is no stage at which this process is complete. One obstacle to the further increase of understanding is the belief that one already understands fully.

We can also see that the deep-rooted conviction mentioned earlier, that it matters whether or not we understand something, is well-founded. For this subjective feeling that we understand something, open to error though it may be, is in general a sign that we are therefore now able to behave appropriately in a new class of situations.

---

[2]I mean here relational understanding. See Chapter 12.

The difference in adaptability between that based on a rule and that which results from understanding has been well demonstrated experimentally by Bell (1967). The example was chosen from topology, a branch of mathematics which will perhaps be new to readers, who may, if they wish, try it for themselves. It has the advantage that the relevant schema can be quickly built up, whereas most mathematical ones take longer.

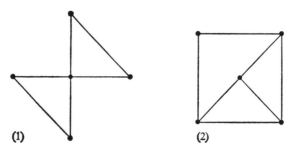

(1)                          (2)

These two diagrams represent topological networks, which are made up of points called *vertices* joined by straight or curved lines called *arcs*. To *traverse* a network means to follow a continuous path covering every arc of the network once and only once. A few trials will show that network (1) can be traversed, whereas (2) cannot. Here are two more examples.

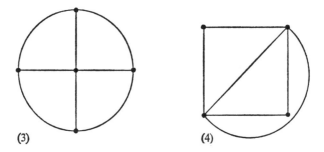

(3)                          (4)

By trial and error, it is easy to find that network (4) can be traversed, and the reader will soon become convinced that (3) cannot, though this is not the same as proving that it is impossible.

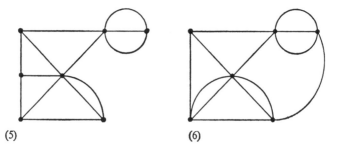

(5)                          (6)

As the networks become more complex, the trial-and-error method becomes more laborious, and its conclusions, particularly if negative, carry less conviction. There is, however, a simple rule. For each vertex, count how many arcs there are which meet there. Call this number the *order* of the vertex. For short we say that a vertex is odd or even according to whether its order is odd or even.

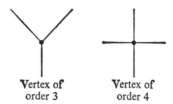

Vertex of    Vertex of
order 3      order 4

*Rule:* a network can be traversed if, and only if, the number of odd vertices is zero or two.

With this rule it is easy to verify that network (6) can be traversed, starting in the top left-hand corner, and that (5) cannot. More complicated networks present little greater difficulty.

Two groups of eleven-year-old children were introduced to the above ideas. Group 1 was given the rule and also an explanation (which will be withheld from the reader at this stage) of the reason for the rule. Group 2 was given just the rule. Both groups of children were then given twelve problems of this kind, including some quite complicated networks. All children of both groups got all the problems right. At this stage, one could not distinguish by their results between children who understood the reason for the rule and those who did not.

A further set of network problems was then presented to these two groups, with one small difference. Here are four typical networks from the set.

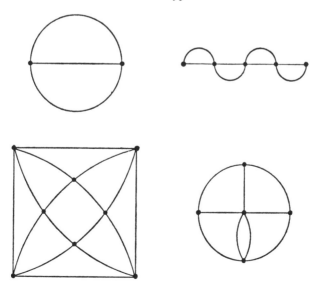

The new problem was (a) to try to find which networks could be traversed as before, but ending this time at the starting point; and (b) to try to find a rule for doing this. Before reading further, readers may care to try for themselves.

A third group of children, with no previous experience of these problems and no knowledge of the rule, was also given this new task. The results, in terms of children finding the correct new rule, were:

| *Group 1* (first rule with understanding) | Nine children out of twelve | (75%) |
|---|---|---|
| *Group 2* (first rule without understanding) | three children out of ten | (30%) |
| *Group 3* (no previous knowledge) | two children out of twelve | (17%) |

Whereas the earlier results of Groups 1 and 2 had been indistinguishable, these new problems show a great gap between them. 75% of the first group were able to adapt to the new task, but only 30% of the second, who did little better than the group with no previous experience.

Now take a sheet of plain paper and copy on it the vertices only of network (1), page 30. Next, draw the network starting at any vertex without lifting the point of your pencil from the paper. (This corresponds to traversing.) Notice that each time you enter and leave a vertex, you add two arcs to the number which meet there, that is, you increase the order of that vertex by two. Now do the same for network (4) and for network (6), starting in the top left-hand corner.

This explanation, which is, of course, briefer than that given to the children, will, it is hoped, provide a sufficient clue for the reader to understand the first rule, given on page 31. If you succeeded in finding the second rule without this explanation, congratulations! If not, it should now be easier.

In the days when so-called teaching-machines were being marketed, I came across an expensive programme called 'Introduction to Topology', published for use with an expensive teaching-machine, in which the first rule (only) was given, and without explanation. In this form, it is not only hard to adapt to problems of the second kind; it does not enable one to answer other relevant questions like 'How can we be sure that this rule works for *any* network?', 'Would it would for a three-dimensional network?' and especially 'How can we be sure that a given network *cannot* be traversed by somebody clever enough? All these questions can be answered by someone who has understood the explanation of the rule, thereby demonstrating further the greater adaptability of the schema to new problems.

## IMPLICATIONS FOR THE LEARNING
## OF MATHEMATICS

The central importance of the schema as a tool of learning means that inappropriate early schemas will make the assimilation of later ideas much more difficult, perhaps impossible.

'Inappropriate' also includes non-existent. Learning to manipulate symbols in such a way as to obtain the approved answer may be very hard to distinguish in its early stages from conceptual learning. The learner cannot distinguish between the two if he has no experience of genuinely understanding mathematics. And all the teachers can see (or hear) are the symbols. Not being thought-readers, they have no direct knowledge of whether or not the right concepts, or any at all, are attached. The way to find out is to test the adaptability of the learner to new, though mathematically related, situations. Mechanical computation does not do this. The amount which a bright child can memorize is remarkable, and the appearance of learning mathematics may be maintained until a level is reached at which only true conceptual learning is adequate to the situation. At this stage the learner tries to master the new tasks by the only means known—memorizing the rule for each kind of problem. This task being now impossible, even the outward appearance of progress ceases, and, with accompanying distress, another pupil falls by the wayside.

An appropriate schema means one which takes into account the long-term learning task and not just the immediate one. The solution of equations, for

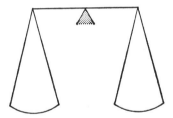

example, is sometimes based on the idea of a pair of scales. If we add or subtract the same weights in each pan, the balance is preserved, and thus we can find a weight which exactly balances the unknown weight. This model also justifies 'taking a number to the other side and changing the sign,' since we get the same result from adding, say, 3 kg. to the left-hand pan, or taking it away from the right-hand pan.

At the beginning stages, this simple schema is admirable. It does have the disadvantage of regarding $x$ as an unknown quantity which we have to 'find,' which is not a basic concept of mathematics, instead of as a variable, which is. But its chief defect is that the schema of 'balancing the two sides' does not apply to equations like

$$x + 4 = 0$$
$$x^2 = 4$$
and $\qquad x^2 - 3x = 4$

except by stretching it till it creaks, and by no device to

$$x^2 + 4 = 0$$
and $\qquad \dfrac{dy}{dx} = 4$

The teacher must look far beyond the present task of the the learner, and wherever possible communicate new ideas in such a way that appropriate *long-term* schemas are formed.

In spite of its shortcomings, the above schema is still incomparably better than the collection of rules without reasons which are sometimes taught, since it does make sense and therefore contributes to an overall belief in mathematics as a meaningful activity. It may also be difficult sometimes to choose between an easy but short-term initial schema and a harder long-term one. This is not quite the same kind of choice as that which we may face when shopping, between something cheap but short-lasting and something more expensive but longer-lasting, since we cannot throw away our early schemas. We have to reconstruct them, which, as we have seen, may present difficulties. So the choice may not always be an easy one. In general, however, it often happens that the more general, long-term ideas are not necessarily harder to learn but only harder to find initially. This transfers the difficulty from the learner to the teacher.

The responsibility of teachers in the early stages of learning is therefore great. They have to make sure that schematic learning, not just memorizing the manipulations of symbols, is taking place. They have to know which stages require only straightforward assimilation and when reconstruction is needed, since at the latter stages the pace must be slower and progress more carefully checked. And they have to plan on a long-term basis the schemas which will be most adaptable to future as well as present needs.

To satisfy fully the latter requirement is impossible. The present rate of change in mathematics, and the uses to which it is put, is such that no one can know what future tasks the present learners of mathematics will have to face. And the rate of change is increasing. So what is best to do?

The first part of the answer would seem to be to try to lay a well-structured foundation of basic mathematical ideas on which the learner can build in whatever future direction becomes necessary: that is, to find for oneself, and help one's pupils to find, the basic patterns; secondly, to teach them always to be looking for these for themselves; and thirdly, to teach them always to be ready to reconstruct their schemas—to appreciate the value of these as working tools, but always to be willing to replace them by better ones. The first of these is teaching mathematics; the second and third are teaching children to learn mathematics. Only these last two prepare children for an unknown future.

# 4

# Intuitive
# and Reflective
# Intelligence[1]

There is an anecdote about a very well-known professor of mathematics which, if it is not true, deserves to be. It relates that, while addressing a learned audience, he wrote a mathematical statement on the board, saying 'This, of course, is obvious.' Looking at it again, he said 'At least, I think it is obvious.' Growing more doubtful, he said 'Excuse me,' and taking pencil and paper, was absent from the room for about twenty minutes. He returned beaming and said triumphantly 'Yes, gentlemen, it *is* obvious.'

Psychologically, the charm of this story is that there is no inconsistency between the first confident statement and the relatively lengthy period of deliberation needed, once doubt had arisen, before this confidence could be regained. By the first statement the speaker meant 'We can accept intuitively the truth of this statement.' By the second statement he meant that, having subjected it to logical analysis, he had confirmed that this intuitive acceptance was justified. Being sure of something is one thing; knowing why one is sure is another.

A similar example. Multiply 16 by 25. (i) What is the answer? (ii) Now explain how you did it. To answer the second question involves turning your attention from the task itself to your mental processes involved in doing it.

Another example. 'What I am writing with is chalk.' 'Chalk is white.' In these two sentences, is the word 'is' used (i) correctly? (ii) with the same

---

[1]Since this book was first published, the term *meta-cognition* has come into use with a meaning which appears to be the same as reflective intelligence. However, I still prefer the latter term, which I took from Piaget (see Skemp, 1961). As a verb, *reflect* comes more easily than *meta-cognise*. Moreover, we do more than cognise our own mental processes. We act on them in various ways, some of which are described in this chapter and in Chapter 8.

meaning? The first question can be answered immediately, but to answer the second question we have to reflect on our use of the word 'is' in each sentence. In these three examples the contrast is between two modes of functioning of intelligence: the intuitive and the reflective. At the intuitive level we are aware through our receptors (particularly vision and hearing) of data from the external environment, this data being automatically classified and related to other data by the conceptual structures described in Chapters 1 and 2. We also act on the external environment by the use of our voluntary muscles acting on our skeleton (a description which includes speech and writing). This activity is largely controlled and directed by feed-back of further information about its progress and result, again via our external receptors. In many cases, it can be entirely successful without any awareness of the intervening mental processes involved, for example, when reading aloud, driving a car or answering the question '6 + 5?'

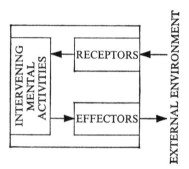

At the reflective level these intervening mental activities become the object of introspective awareness. A child asks us why we pronounce the word 'accelerate' as 'axelerate' not 'ackelerate.' So we explain (in terms appropriate to the hearer and with examples) that the first c is hard, because followed by a consonant, the second soft, because followed by e or i. Our pronunciation is explained by showing it to be consistent with certain accepted classes of response. Or, a learner-driver asks us why we changed gear before reaching a sharp bend in the road. Though we had done so 'without thinking' (that is to say without reflection), we have no difficulty in explaining our reason. Or, having replied '400' to the question '16 × 25,' we might be asked 'How did you do that in your head?' And having described our method (there are several to choose from), we might also be asked to justify it—a much more searching question, as the answer involves reference to the associative property of multiplication.

The data necessary to answer all of these questions comes not from the environment but from our own conceptual systems. These are represented in the diagram on page 37 by the block labelled 'intervening mental activities.' We direct our attention to this source of data so easily and habitually that we take for granted this ability to reflect on our own mental processes. But we should be much more surprised at it than we are. Our awareness of the outside world can be accounted for by obvious sense organs—eyes, ears, etc.—and the neural paths from these are traceable. But no neuroanatomist has yet revealed the internal equivalent by which we can 'see' our own visual imagery or 'hear' our internalized speech.

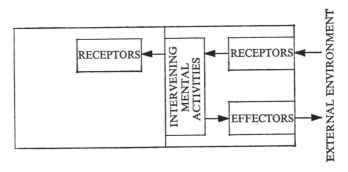

Moreover, this ability is much less developed in young children. Here are two examples from the work of Piaget (1928).

Weng (age 7): 'This table is 4 metres long. This one is three times as long. How many metres long is it?' '*12 metres.*' 'How did you do that?' '*I added 2 and 2 and 2 and 2 and 2, always 2.*' 'Why 2?' '*So as not to take another number.*'

Gath (age 7): 'You are 3 little boys and are given 9 apples. How many will you each have?' '*3 each.*' 'How did you do that?' '*I tried to think.*' 'What?' '*I tried to think in my head.*' 'What did you say in your head?' '*I counted . . . I tried to see how much it was and I found 3.*'

Being able to do something is one thing; knowing how one does it is quite another. There are, however, considerable individual differences in this, and the writer recently obtained the following replies from a child aged six years ten months. (To the first question, with feet instead of metres.) '*12 feet.*' 'Can you say how you found the answer?' '*Well I went 3, 6, 9, 12.*' (To the second question.) '*Three.*' 'How did you find out?' '*3 and 3 and 3 make 9.*' (Then, a spontaneous afterthought.) '*The quickest way is to write [sic] 3 times 3 is 9.*'

Once we have become able to reflect, to some degree, on our own schemas and how we use them, important further steps can be taken. We can communicate these, as in the foregoing example. We can set up new schemas and make new plans based on these. Someone unable to do the example cited earlier (16 × 25) might, after it had been pointed out that four twenty-fives make a hundred, not only be able to work out 16 × 25 by thinking of it as 4 × (4 × 25) which is equal to 4 × 100, but also work out other multiplications, like 24 × 25 and even 25 × 25. If the person could do all these, it would indicate that a simple schema and not just the answer to a particular question had been acquired.

We can replace old schemas by new ones. If readers have tried to back a car with a boat trailer or caravan attached, they may appreciate the following non-mathematical example. The writer had been told to 'put the steering wheel down on the side you want the trailer to go'. This was not very successful, however, and his wife suggested the following alternative approach. 'If you were just pushing it by hand, you would have no trouble in steering it, would you? So imagine yourself pushing on the towing hitch, but using the car to push with.' Substitution of this schema proved strikingly successful, since backing the car itself in any desired way was already automatic.

We can correct errors in existing schemas. If we say 'I see what I was doing wrong,' this implies not only a reflection on our existing method but the discovery of the particular details in it which were causing failure, followed, usually, by a deliberate change in these details.

Just how we are able to make deliberate changes in our schemas, as a whole or in detail, is still unknown. Since, however, we can certainly do so, our diagram needs further additions.

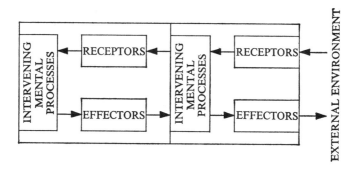

Here are some further examples which involve reflective activity.

Someone wants to know how to multiply together two decimal fractions, say, 1·2 and 0·57. So we explain to them how the decimal point may be omitted, the multiplication done in the ordinary way and the decimal point then re-inserted by counting the total number of figures after the decimal point. ($12 \times 57 = 684$; 1·2 has one figure after the decimal point, 0·57 has two—total three; re-insert the decimal point in the result to give three figures after the decimal point—result 0·684.) This rule will enable them to get the right answer, but it is unrelated to their existing knowledge of the meaning of decimal notation. To explain the method, we could rewrite the decimals as common fractions:

$$1.2 \times 0.57 = \frac{12}{10} \times \frac{57}{100} = \frac{684}{1000} = 0.684$$

The power of 10 in the denominator = the number of zeros in the denominator = the number of places after the decimal point. Multiplying the denominators corresponds to adding the numbers of zeros, which corresponds to adding the numbers of decimal places.

Having done all this, we could go further and reflect on our method of communication. We might then decide that it would be better next time to demonstrate the meaningful method first, before showing (or encouraging the learner to seek) the short cut. So we would reorganize our plan for communicating the schemas for multiplying decimals.

A far-reaching kind of reflective activity is that which leads to mathematical generalization. In the process of learning the use of indices, for example, we can distinguish two distinct stages. After defining the notation, by examples such as

$$a^2 = a \times a \qquad \text{(where } a \text{ is any number)}$$
$$a^3 = a \times a \times a$$
$$a^4 = a \times a \times a \times a, \qquad \text{etc.}$$

it is easily seen that

$$a^2 \times a^3 = a \times a \qquad \times \qquad a \times a \times a$$
$$= a^5$$

and from this and similar examples learners form, intuitively, the general schema whereby they can write directly

$$a^5 \times a^7 = a^{12}, \qquad \text{etc.}$$

By using the methods for manipulating algebraic fractions already known to them, they can also form a schema for division, derived from examples such as

$$a^5 \div a^2 = \frac{a \times a \times a \times a \times a}{a \times a} = a \times a \times a = a^3$$

whereby they can write directly

$$a^{15} \div a^6 = a^9, \qquad \text{etc.}$$

Having formed these two (related) schemas, they can also *formulate* them: that is, express them symbolically in the form

$$a^m \times a^n = a^{m+n}$$
$$a^m \div a^n = a^{m-n}$$

where $m$ and $n$ stand for any two natural[2] numbers other than zero, and in the second case $m$ is greater than $n$. These formulations detach the methods from any particular example of their use and enable them to be examined as entities in themselves. The restrictions that $m$ and $n$ must be natural numbers and $m$ greater

---

[2]That is, counting-numbers such as 1, 2, 3, etc.

than $n$ were made necessary by the initial definition of $a^2$, $a^3$ . . ., since symbols like $a^0$, $a^{-2}$, $a^{1/2}$, have no meaning in terms of this definition. But the methods have now become partially detached from their origins, and restrictions which at first seemed right and proper now become open to question. Under what conditions is it (1) permissible and (2) advantageous to remove these restrictions?

A reasonable criterion for the first is that the new method shall not introduce any inconsistency with known methods; and for the second, that by removing the original restrictions the advantages of index notations can be usefully and meaningfully extended.

Many readers will be familiar with the extensions of index notation whereby

$a^0$ is given the meaning $\quad 1$

$a^{-2}$ is given the meaning $\quad \dfrac{1}{a^2}$

$a^{\frac{1}{2}}$ is given the meaning $\quad \sqrt{a}$

and so on. With these and similar meanings for negative and fractional indices, the original restrictions can be removed. We say that the notation and method have been *generalized.*

What are the mental processes involved?

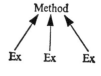

From a set of examples, a general method is derived, which can be applied to other examples *of the same kind.*

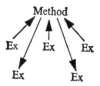

The method is next formulated explicitly, considered as an entity in itself, and its structure analysed.

This structure is used to invent ways of using the same method for examples of a new kind. The original examples are included in the enlarged field of application of the method.

This process of mathematical generalization described above is a sophisticated and powerful activity. Sophisticated, because it involves reflecting on the *form* of the method while temporarily ignoring its content. Powerful, because it

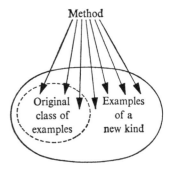

makes possible conscious, controlled and accurate reconstruction of one's existing schemas—not only in response to the demands for assimilation of new situations as they are encountered *but ahead of these demands,* seeking or creating new examples to fit the enlarged concept. Trying to do this intuitively is a much more hit-or-miss affair and will not perform to order. It must be accepted that the intuitive leap is a frequent forerunner of the deliberate generalization, suggesting a direction which might otherwise have remained unexplored. But intuition sometimes 'lets one down.' That is, when subjected to critical analysis, weaknesses are found—inconsistencies with accepted ideas, which make true assimilation to existing (and well-tried) principles impossible. The learned professor mentioned at the beginning of the chapter was right to be cautious of his intuitive judgement until he had tested it analytically.

A widely encountered example of successive mathematical generalization is that of number. Both historically and for the individual learner, the natural numbers come first. These are properties of sets of discrete (and so countable) objects, and methods for adding and subtracting, multiplying and dividing these, developed over the centuries, are learnt in their first decade or so by children of our own culture. Subsequently other things are encountered called 'fractions' and 'negative numbers,' and rules are given which are alleged to be the correct way to add and subtract, multiply and divide, these. Correct according to what criterion? All too frequently the only one offered is whether the teacher decides that the rules have or have not been correctly followed. In such cases understanding has been lost, perhaps never to be recaptured. Worse, 'making sense' has ceased to be the criterion by which a mathematical statement is judged. Worst, another learner has been convinced that mathematics is boring and meaningless—true of what is presented under this guise, but false of mathematics.

How may the idea of number be successfully generalized through the stages of fractional numbers, integers, rational numbers, etc.? Briefly, we need to formulate what are the formal properties of the system of natural numbers. By the *system* of natural numbers we mean the set of natural (counting) numbers together with the operations of addition and multiplication, whereby any two members of the set can be combined (in one way or the other) to get another member of the set. By the formal properties we mean those properties which do not depend on the particular examples we choose. So $12 + 9 = 21$ and $12 \times 9 = 108$ are not formal properties, but $12 + 9 = 9 + 12$ and $12 \times 9 = 9 \times 12$ are, though not stated in a general way. The five formal properties of the natural-number system are:

$$a + b = b + a$$
$$a \times b = b \times a$$
$$a + (b + c) = (a + b) + c$$
$$a \times (b \times c) = (a \times b) \times c$$
$$a \times (b + c) = a \times b + a \times c$$

where $a$, $b$, $c$, are any (natural) numbers. It is tempting to regard these properties as trivial, but they are the very foundation of all numerical manipulation. For example, without the first property the size of our shopping bill would depend on which we bought first; and without the third, it would depend on which two items the assistant added together first. These five properties are also, with the help of index notation, the foundation of elementary algebra.

Invaluable though our system of counting numbers is, it has its limitations. With the help of units it can be extended to make possible the measurement of continuous objects, but we soon find that the existing numbers do not include all we need to deal with sizes less than a unit. So new numbers corresponding to these broken units are introduced. But we are premature in calling them numbers—before we generalize the 'number system' schema, we must satisfy the two requirements of consistency and usefulness. (A pure mathematician would be content with the first, but the second usually follows, sooner or later. Few mathematical ideas of great generality fail to be useful in the course of time.)

Consistency means that we have to invent ways of 'adding' and 'multiplying' these new entities which have the five formal properties already listed. Usefulness means that the results of these manipulations must tell us something we want to know in terms of the material objects to which these entities refer. And though this is not an essential, it will be a great help if the signs for these new entities can be developed out of signs in common use already (just as we use the letters of our existing alphabet for newly invented words), and if the methods for 'adding' and 'multiplying' can make use of the large number of addition and multiplication results which we have already learnt. All these requirements, when satisfied, make possible the assimilation of the new number system to our existing and well-practised schema.

The way in which they are all met is a fascinating subject, and the reader who explores it further will learn much about the foundations of mathematical thinking. The same applies to the development of positive and negative integers, rational numbers (often identified with fractional numbers) and real numbers (the system which includes irrationals like $\sqrt{2}$, $\pi$). Here we are concerned mainly with the process rather than the result, and particularly with the activity of reflecting on the formal properties of the schemas, which is part of the process of mathematical generalization and one of the most advanced activities of reflective intelligence.

If this second-order functioning of intelligence is of such importance for progress to the more advanced levels of mathematics, it is clearly of great interest to know at what ages it makes its appearance, and how (if at all) we can aid and perhaps hasten its appearance. In answer to the first question, we have a body of research by Inhelder and Piaget (1958) which indicates that children develop the ability to reflect on *content* during the ages from about seven to eleven and to manipulate concrete ideas in various ways, such as reversing (in imagination) an action, thereby returning to the previous state of affairs. They found, however, that their subjects could not reason formally—consider the form of an argument independently of its content—until adolescence. Closely related to this, they found that the younger children could not argue from a hypothesis if this hypothesis was contrary to their experience.

In this research the subjects were taken 'as they were found'. That is to say, the experiments indicate the progress of reflective intelligence as it developed in the Swiss schoolchildren from whom the subjects were taken, by the interaction of their innate abilities with the cultural and educational experiences which they encountered. What we do not at present know is the extent to which the rate of development might be helped by teaching. To consider a parallel: most children learn to sing spontaneously, just from growing up in a culture where they hear other people singing. But this learning is greatly accelerated, and the final degree of accomplishment greatly raised, in, say, boys who become members of the choir of King's College, Cambridge, or Magdalen College, Oxford. At present the development of reflective ability and formal reasoning is not the subject of deliberate teaching, partly because its importance has hardly been realized and partly because we do not know how to teach it, which presupposes a knowledge of how it is learnt.

A reasonable hypothesis about the latter is that any situation which requires learners to formulate ideas explicitly, and to justify them by showing them to be logically derivable from other and generally accepted ideas, would exercise and so develop the ability to reflect on their schemas. In other words, argument and discussion are useful ways of learning to reflect.

Those who have tried usually agree that trying to teach a topic exerts strong pressure to clarify one's own thinking about it. A simple experiment has given support to this view. Two parallel forms of secondary-school boys, aged about fourteen, were taught different topics by their regular mathematics teachers. Each form was given a test on the topic it had been taught and divided into two halves of equal performance (as nearly as possible) as measured by these tests. One half of each form then taught what they had learnt to their opposite numbers in the other form, while the other (matched) half spent the same time in further practice at the same topic. The boys who were acting as teachers thought that their pupils were going to be tested on what they had been taught by them. Actually, at the end of the experiment all were re-tested on the topic which they had learnt during the first part of the experiment, the aim being to compare the effects of teaching a topic to someone else and continuing to practise it by oneself. The results came out quite clearly in favour of the former.

Communication seems to emerge as one of the influences favourable to the development of reflective intelligence. One of the factors involved is certainly the necessity to link ideas with symbols: this will be considered at length in the next chapter. Another is the interaction of one's own ideas with those of other people. To the extent that agreement is reached, the resulting communality of ideas is less egocentric, more independent of individual experience. And as has already been suggested, the cut and thrust of intellectual discussion forces on one the necessity to clarify ideas in one's own mind, to state them in terms not likely to be misunderstood, to justify them by revealing their relationships with other ideas; and also, to modify them where weaknesses are found by the other side, ending with a stronger and more cohesive structure than before. To be able to do the last requires the achievement of a partial detachment from one's ideas, a state of reduced involvement, so that one does not feel personally attached, injured or defeated when one's schemas are shown to have some inaccuracy or inconsistency. This is yet another aspect of the state of reflection. It is also largely dependent on the interpersonal situation, an aspect which will be discussed in Chapter 7. The last consideration suggests that relationships with teachers may be of great long-term importance in the development of reflective intelligence. But on this, there is as yet no evidence.

A caveat is desirable here. The preceding discussion has to some extent carried the implication that an individual is 'at the intuitive stage,' 'capable of reflection on form and content combined,' 'capable of formal reasoning,' in general; that is, if he is at a given stage relative to topic A, he is at the same stage relative to topic B. But it may well be the case that we all have to go, perhaps more rapidly than the growing child, through similar stages in each new topic which we encounter—that the mode of thinking available is partly a function of the degree to which the concepts have been developed in the primary system. One can hardly be expected to reflect on concepts which have not yet been formed, however well-developed one's reflective system. So the 'intuitive before reflective' order may be partially true for each new field of mathematical study. Further research is needed here.

While learners are still at the intuitive stage they are largely dependent on the way material is presented to them. If the new concepts encountered are too far removed from any of their existing schemas, they may be unable to assimilate them, particularly if reconstruction is required, for this is largely dependent on reflection. So in the earlier stages a conceptual analysis by the teacher must be used as a basis for a careful plan of presentation, from which learners can re-synthesize the structures in their own mind. This is the case whether learning takes place directly from a live teacher, or indirectly from a book. The former situation has the advantage that questions can be asked, explanations given; an even greater advantage is that a sensitive teacher can perceive the growing points of the learner's schema and offer the right material at the right moment. This flexibility of approach entails also a greater mastery of the subject than keeping strictly to a prepared plan, however good.

The final contribution of the excellent teacher is, however, gradually to reduce the learner's dependence. When my young son was first doing jigsaw puzzles, his mother or I used to offer him pieces which fitted on to what he had already put together. But a stage was reached when he did not like us to do this any more, and it is towards this kind of independence that the mathematics teachers must work. Once people are able to analyse new material for themselves, they can fit it on their own schemas in the ways most meaningful to themselves, which may or may not be the same ways as it was presented.

So the teacher of mathematics has a triple task: to fit the mathematical material to the state of development of the learners' mathematical schemas; to also fit the manner of presentation to the modes of thinking (intuitive and concrete reasoning only, or intuitive, concrete reasoning and also formal thinking) of which the learners are capable; and, finally, to increase gradually the learners' analytic abilities to the stage at which they no longer depend on their teacher to predigest the material for them.

And although we have some reasonable conjectures about how this last development may be encouraged, our knowledge in this area is far from complete. In this respect, as in many others, the best teachers are those who are still active learners.

# Symbols

In previous chapters we have considered the formation of concepts, the function of schemas (conceptual structures) in integrating existing knowledge and assimilating new knowledge, and the additional power which comes from the ability to reflect on one's schemas. In each of these processes an essential part is played by symbols, which have other functions as well. It is now time to consider these in detail.

Among the functions of symbols, we can distinguish:

    (i)  Communication
    (ii)  Recording knowledge
    (iii)  The communication of new concepts
    (iv)  Making multiple classification straightforward
    (v)  Explanations
    (vi)  Making possible reflective activity
    (vii)  Helping to show structure
    (viii)  Making routine manipulations automatic
    (ix)  Recovering information and understanding
    (x)  Creative mental activity

Most of these are related, particularly to the first. Recording knowledge is communicating with the reader, explanation is a special kind of communication, reflecting is communicating within oneself; and other connections will also be apparent. The headings are therefore intended for convenience only, as starting points for the discussions which follow, not as partitions.

## COMMUNICATION

A concept is purely mental object—inaudible and invisible. Since we have no way of observing directly the contents of someone else's mind, nor of allowing others access to one's own, have to use means which are either audible or visible—spoken words or other sounds, written words or other marks on paper (notations). A symbol is a sound, or something visible, mentally connected to an idea. This idea is the *meaning* of the symbol. Without an idea attached, a symbol is empty, meaningless.

Provided that a symbol is connected to the same concept in the minds of two people, then by uttering[1] this symbol, one can evoke the concept from the other's memory into their consciousness—can cause them to 'think of' this concept in the present. This proviso is, however, no small one. Once the connection is established, its meaning is projected on to the symbol, and the two are perceived as a unity. So it is hard to realize that what is meaningful to oneself may not be meaningful to the hearer—a difficulty experienced by many when speaking to foreigners—or that the same meaning is not being attached, for example, to the word 'braces', which may mean to someone British a device for holding up one's trousers, but to an American a pair of set brackets { }. We may think that we are communicating when we are not, and, indeed, it is impossible to know for certain whether we are, and, if so, to what degree. For the reason given above, we usually take it for granted, but the communication links are so precarious, and so inaccessible to study, that we would do better to be surprised that we can communicate our ideas to each other at all. After all, it has taken millions of years of evolution to produce an animal which can do so to any marked extent.

Let us take as a starting point (a) that a symbol and the associated concept are two different things; (b) that this distinction is non-trivial, being that between an object and the name of that object. If an object is called by another name, we do not change the object itself, and this is still true for an object of thought—in the present context, a mathematical idea. For example,

$$\text{'five,' 'cinq,' '5,' 'V,' '101'}$$

all refer to the same number in different notations. We do not call five an English number and *cinq* a French number, nor should we call 5 an Arabic number and V a Roman number. But we still read, all too often, instructions to pupils like "Turn the binary number 101 into a decimal number.' The whole object is, of

---

[1]This will be used as a convenient condensation for speaking, writing, drawing, projecting on a screen, etc.

course, *not* to change the number itself in the process of representing it in a different way. In translating from French into English, we try to keep the meaning the same while changing the words. In converting pounds to dollars we try to keep the value in goods or services the same while representing this value by different tokens (coins, notes) or symbols (figures on a cheque or bank transfer).

The term 'binary number' also implies that being binary is a property which a number can have or not, like being even, prime, an integer, etc. But binary *numerals* can be used to represent any kind of number at all, odd or even, prime or factorizable, natural number, integer, rational or real number. One of the first requirements of communicating an idea is to be clear about it oneself. Those who talk or write about 'binary numbers' and 'decimal numbers' are not.

Usually, when uttering a symbol, we want to call to the attention of the receiver the idea attached to the symbol rather than the symbol itself. If it is the symbol we are referring to, we can show this by quotation marks. (More symbols! They are inescapable.) Example:

'5' and 'V' are both symbols for (the number) five.

A symbol for a number is called a 'numeral', and a system of numeration is a system for writing as many different numbers as we like with a relatively small number of digits (single numerals like 0, 1, 2, 3, 4 . . . 9). The decimal system uses ten digits, the binary system uses two. If it is not clear from the context which system is in use, this can be shown simply and clearly by a suffix. The sign = ('is equal to') means that we are referring to the same concept, (usually) by different symbols. So, for example,

$$5_{ten} \quad = \quad 101_{two} \quad \text{(since 101 in binary means the same as 5 in decimal notation.)}$$

Similarly $8_{ten} \quad = \quad 10_{eight} \quad = \quad 1000_{two}$ etc.
But $\quad$ '$8_{ten}$' $\neq$ '$10_{eight}$.' The numbers are equal, the numerals are different

Excessive precision in the use of language[2] is rightly regarded as pedantry. So it is a fair question at this stage to ask whether this label is applicable to the preceding discussion. Does it really matter, for example, which of these we say or write:

'Write the binary number 11010 as a decimal number' or
'Write $11010_{two}$ in decimal notation'?

An easy defence would be to claim that it is part of the duty of a mathematician to be as accurate as possible all the time. But this, though plausible, is not valid. It would, for example, imply that we should never use convenient but

---

[2]A symbol system; for example, the English language, the language of mathematics.

loose phrases such as 'as small as we like.' Part of the aim of mathematics is, by abstraction and the omission of irrelevancies, to enable us 'to see the wood for the trees,' and this will not be achieved by adding, instead, a mass of mathematical detail in the name of accuracy.

The kind of accuracy with which we are at present concerned is accuracy of communication, with trying to get as near as we can to the impossibility of producing the same idea in the mind of the receivers as of the communicators or calling it to their attention.

Now, we can distinguish three categories of hearer or reader. First, those who don't yet know what we are talking about, but want to. For these, we should choose our symbols with the greatest possible care and use them as accurately as we can, with the aim of communicating nothing but the truth, though not yet necessarily the whole truth. Concepts are built up by degrees. The first approximation is bound to be incomplete, and perhaps to need tidying up in detail, but there should not be anything of importance to un-learn. It is also worth bearing in mind that, to an intelligent learner, a brief but inaccurate statement may well be more confusing than a somewhat lengthier, but accurate, statement.

The second category comprises those who do know what we are talking about, as a general background within which we are trying to communicate some particular aspect. If they are willing to 'go along' with us, we can take much for granted, save time and concentrate on essentials. An old and wise teacher of mine often used, in the context of limits and convergence, phrases like 'As near as dammit to . . .' We both knew what he meant, and both could, if necessary, have re-phrased it in rigorous terms. So, for the task in hand, the idea was communicated with complete accuracy by this short and expressive phrase.

The third category of hearer or reader consists of those who do know what we are talking about but want to fault it. A non-mathematical example of this activity is to be found every time a new tax is made law. The finance minister says 'I want a tax on . . .' As soon as this becomes law, an army of expert accountants will go to work on behalf of their clients to see how this tax can legally be avoided or reduced. So, before the bill goes through, the parliamentary draughtsmen have to try to stop all loopholes in advance. The result is to make it almost unintelligible.

Similarly in mathematics, rigour and ease of understanding do not go together. The art of communication is, first, to convey meaning. Afterwards, the new ideas can be subjected to the stress of analysis, and greater precision introduced where weaknesses are found. The difference is that, once a schema is well established, this critical attack serves a useful purpose, that of stimulating more careful formulation and greater reflective awareness, and the strengthening of the schema without loss of integration of 'the overall picture.' This criticism may come from another person or it may come from a 'devil's advocate' within oneself. This seems to be another function of the reflective system—to take 'an outside view' of an argument or other intended communication, and by self-criticism anticipate external criticism.

## RECORDING KNOWLEDGE

Ideas are not only invisible and inaudible, they are perishable. When we die, our knowledge dies with us, unless we have communicated or recorded it. One of the most moving episodes in the history of mathematics is that of the young Galois (1811–32) sitting up all night, writing against time to commit to paper his theory of groups, before his tragic and wasteful death by duel at the age of twenty.

Recording is a special case of communicating, since it is normally done with the intention that these records will, in the near or distant future, be seen by others. So all the previous section applies. Whereas the spoken communication usually (though not always) takes place in circumstances which allow questions and explanations to be given, written or printed symbols have to convey all the required meaning, without a second chance on either side. So the communicators have to take more trouble to try to ensure this. There is, however, the advantage that the receivers have a permanent record for revision and the checking of earlier points. They can also go at a speed to suit their own rate of assimilation.

As has been discussed in Chapter 1, the conceptual structure of mathematics is something far beyond that which anyone could construct, unaided, in a lifetime. Limited areas have taken years of work by some of the world's most gifted individuals. It is the storage of the accumulated knowledge of previous generations by written and printed symbol systems (and recently by other techniques such as recording tape, cinematography, microfilm), together with the auxiliary explanations of live teachers, that make it possible for some of each new generation to learn in years ideas which took centuries of collective effort to form for the first time, in each case, synthesizing them anew, and in some cases building new knowledge and adding this to the store.

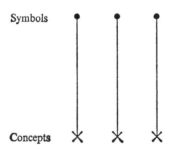

One of the first requirements for the avoidance of ambiguity which one would expect to be observed is that each symbol is associated with one concept, and vice versa. This arrangement is, however, seldom found in practice, even in a single language. Mathematicians seem to be particularly lazy about inventing new symbols, relying largely on the capital and lower-case letters of the Roman alphabet, the Greek alphabet, punctuation marks and the like, each of which does multiple duty. So a single symbol may well stand for a variety of concepts.

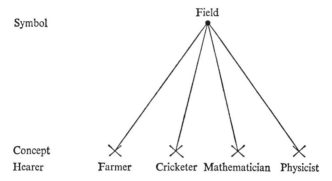

Symbol

Field

Concept
Hearer          Farmer      Cricketer  Mathematician  Physicist

The arrangement just shown might be expected to lead to confusion, since the word 'field' will evoke different concepts in the minds of each of the individuals named above. Or, if we are addressing someone with interests in all these topics, then we cannot be sure which concept will be evoked by the word 'field' in isolation. But, of course, the word is seldom used in isolation. Ordinarily the

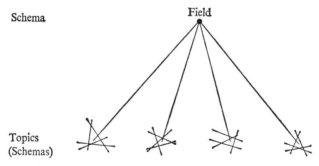

Schema

Field

Topics
(Schemas)

hearer knows which topic is under discussion, and only ideas within this topic are accepted as possible meanings for the word. If not, then the speaker or writer uses one or more other symbols to evoke the relevant schema as a whole. This established a 'set'—a state of mind in which concepts belonging to this particular schema are more easily evoked. Symbols used in this way, to determine the schema within which a particular symbol takes its meaning, are called its *context*.

From this, three simple rules can be formulated for conveying the desired meaning when one symbol corresponds to many concepts.

1. Be sure that the schema in use is known to the hearer or reader.
2. Within this schema let each symbol represent only one idea.
3. Do not change schemas without the knowledge of the hearer or reader.

It is permissible (though whether any advantage is gained is another question) to use the same symbol in different contexts with different meanings. But in the same context a symbol must have just one meaning. So we can write $AA' = I$ in the context of matrices, and $AA' = BB'$ in the context of points and lines, without confusion. But if we write $(x + a)^2 = x^2 + 2ax + a^2$ the $x$ and the $a$ must keep the same meaning throughout, because they are in a single context.

These rules seem straightforward and obvious, but they are not always observed, with the result that the learner is confused. Here is an example.

Children first learn the meaning of multiplying in the context of natural numbers, which refer to sets of discrete, countable objects. So the operations $3 \times 4$ corresponds to combining four sets, each of three objects, and counting the objects in the resulting set.[3] They use the sign '$\times$' with this meaning for several years, and it is the only meaning they know. We then change to a new number system, say, fractional numbers or integers, in which the sign (or word) has a different meaning. But we do not tell the children that we have changed the context and have generalized the meaning of '$\times$' to suit the new context. So they no longer fully understand what they are doing.

If the new context was very different from the old, children would probably discover what was happening unaided. But the contexts are sufficiently alike to make it hard for them to do so. One way of indicating the change is already in use in advanced texts. The symbol '$\otimes$' (and also '$\oplus$') is used in the new context, to show that these operations are like the others but that we must not expect them to be quite the same. The readers of these texts probably come into the third of the categories outlined on page 49, those who will be quick to notice any inaccuracy. But those for whom accuracy of communication is most necessary are those in the first category, those who do not yet know what we are talking about but want to. When these pass on into category two, we may conveniently revert to the symbols '$+$' and '$\times$', since they are now able to assign the appropriate meanings according to context.

The word 'line' is commonly used with at least three different meanings: (a) a line indefinite length, extending indefinitely in both directions; (b) one which starts at a given point and extends indefinitely in one direction from it; and one which is of finite length, bounded by two points. These three meanings may conveniently be distinguished by the terms 'line,' 'ray' and 'line segment.' So the point $X$ is on the line $AB$ (or $BA$), and also on the ray $BA;$ but it is not on the ray $AB,$ nor on the line segment $AB$. If $AB$ represents a railway line, $X$ our destination and $A$ our starting point, the distinction is hardly trivial!

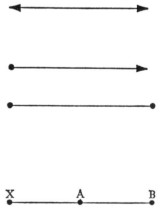

---

[3]That is, assuming that we read '$3 \times 4$' as 'three multiplied by four.' It is also read by some as 'three times four,' which corresponds to combining three sets, each of four objects. We should be more surprised than we are that both of these give the same result.

The mathematically experienced reader should have no difficulty in finding other examples of the ambiguous use of symbols. Some suggestions: what is meant by '$AB$ = 3 cm?' What is meant by 'the series $1 + \frac{1}{2} + \frac{1}{4} + \frac{1}{8} + \frac{1}{16}$ etc.?' And in the context of groups, are the terms 'identity element' and 'neutral element' synonymous?

So far, the emphasis of this section has been that, in a given context (which may be explicit or implicit), one symbol should represent only one concept. What matters is the meaning (the associated concept), and provided that each

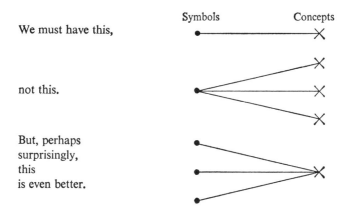

symbol conveys only one meaning, it is often an advantage to have a choice. If A uses a term (for example, 'cuboid') which is unfamiliar to B, they can try again with another (say, 'rectangular block'). The choice of symbol also enables us to classify the same idea in different ways, a use which will be discussed further in section (iv) of this chapter; and, related to this, it can help us to emphasize that aspect of a complex idea which is most relevant to particular circumstances. For example, *function* is a concept with widespread applications, and in Chapter 13 we shall see that there are no less than six useful ways of representing a given function.

Other advantages of using several different symbols for the same concept will be mentioned later in this chapter. If we do this, however, an obvious precaution is necessary to ensure that the reader knows that we are in fact talking about the same thing, though using different names; and this becomes more important when recording mathematics, as distinct from communicating face-to-face, since the reader cannot ask. This is the meaning of the symbol '$=$,' that the symbols on each side of the sign of equality refer to the same object.

## THE COMMUNICATION OF NEW CONCEPTS

It will be recalled that in Chapter 2 the point was made that new concepts of a higher order than those which the learner already has can only be communicated by arranging for the learner to group together mentally a suitable set of examples.

If the new concept is a primary concept, for example, red, it is possible to do this without the use of symbols, simply by pointing. The words 'This is a . . .' simply help to draw attention; they are verbal pointers. 'Red tie,' 'red book,' 'red pencil,' 'red light,' however, express simultaneously the variability of the examples and the constancy of the concept. Intuitively the learner associates the invariant property with the invariant word, and so learns the name for the concept while it is being formed.

If the concept is a secondary concept, as are all mathematical concepts, then the only way of bringing together a suitable set of examples in the learner's mind is to bring together the corresponding words. 'Red, blue, green, yellow—these are all colours.' By manipulating the words we manipulate the minds of the learners—normally, with their consent. (If they feel otherwise, there will naturally be resistance to learning: see Chapter 7.) In this way learners may be helped to see something in common between examples which, separately encountered over an interval of time, would have remained isolated in their minds. It took Newton to perceive for the first time something in common between the fall of an apple and the motion of the planets round the sun; but when he brings these ideas together for us, we too can form the concept of gravitation.

Another way of communicating new concepts is by relating together classes already known to the hearer. 'What is a Sinhalese?' 'An inhabitant of Sri Lanka.' 'What is a kite?' (In the context of geometry.) 'A quadrilateral with two pairs of adjacent sides equal.' 'What is a variable?' 'An unspecified member of a given set.' If the hearer already has the class concepts mentioned, this implies that examples of these are known, so it should also be possible to supply examples of these new concepts. Indeed, this is often the first response, partly to confirm that the concept has been understood. (Sketching rapidly in response to the second definition: 'Like this?')

But the response also seems to satisfy a deeper need. Somehow, a concept acquired in the way just described seems incomplete until it has some examples. A tentative explanation of this is that a concept confers the ability to class together an appropriate set of examples, and it is generally observable that the acquisition of a new ability often seems to carry with it a need to exercise it. (Give your small son a kit of tools for his birthday and observe the result.)

The examples of the new concept thus supplied need not be from past experience. One can imagine a Sinhalese without ever having met one; one can imagine a 100-sided regular polygon without having seen one and without having to draw one. Indeed, a fruitful and exciting method of mathematical generalization is to invent a new class, and then try to find some members of it. Example: suppose that we already have the concepts square root and negative number, and combine these to form a new concept—the square root of a negative number. The search for examples of this new class, and the investigation of their properties, leads to the construction of a new set of ideas which, though termed 'imaginary' numbers, are nevertheless of great practical use in physics: for example, in the theory of alternating current and oscillatory circuits.

## MAKING MULTIPLE CLASSIFICATION STRAIGHTFORWARD

A single object may be classified in many different ways, and, by using different names for it (which we have already seen to be permissible), we can indicate what particular classification is currently in use. The same man may be called 'Mr John Brown,' 'Sir,' 'The right honourable gentleman,' 'Uncle Jack,' 'Daddy,' or 'John.' The same angle may be classified as the angle vertically opposite to . . . or as the third angle of triangle . . . The same number may be regarded as the square of 8, the cube of 4 or the square of 10 minus the square of 6, may be symbolized by $8^2$, $4^3$, $10^2 - 6^2$. By our choice of symbol, we are enabled to concentrate our attention on different properties of the same object.

As already noted, we show that we are still (often in spite of appearances) referring to the same object[4] by the symbol '$=$,' and, by renaming according to already established routines, we can find properties which were at first not apparent.

Example: $4x^2 - 12xy + 9y^2$, where $x$ and $y$ are both numerical variables (unspecified numbers). We know that this collection of symbols represents some number. But by writing

$$4x^2 - 12xy + 9y^2 = (2x - 3y)^2$$

we know something new—that it represents a *positive* number.

Though the principle is a simple one, its consequences are far-reaching. Once we have appropriately classified something, we are a long way towards knowing how to deal with it. (This polite caller—is he a salesman, a public-opinion surveyor or a plain-clothes detective? Our response is cautious until we know which.) 'Appropriately' means in a way (or ways) which helps us to solve the problem in hand; and so the more ways in which we can classify, the greater the

---

[4]Reminder: this, in the present context, usually means an object of thought.

variety of problems which we can solve. And the more symbols we can attach to the same concept, the more ways can we classify.

## EXPLANATIONS

An explanation is a communication intended to enable someone to understand something which they did not understand before. Understanding results from assimilation to an existing schema, so where this has failed there are three possible causes.

(a) The wrong schema may be in use. In this case the explanation needed is simply to activate the appropriate schema. In the present book, words such as 'function,' 'image,' 'group,' are used both in the everyday sense and in the mathematical sense. Failure of understanding could result from attaching a different meaning from that intended. This is simply a matter of context.

(b) The gap between the new idea and the (appropriate) existing schema may be too great. Using again the indices example (page 39), suppose that one began by showing the notation

$$a^2 = a \times a$$
$$a^3 = a \times a \times a$$

and then continued straight to $a^m \times a^n = a^{m+n}$. Very likely the learners would say that they did not understand, perhaps adding 'You have gone too fast.' The explanation needed here would be to supply more intervening steps, thereby bridging the gap. In psychological terms, the explainer would utter suitable symbols by which to evoke concepts relating the existing schema to the new idea.

(c) The existing schema may not be capable of assimilating the new idea without itself undergoing expansion or restructuring, of which a special case is mathematical generalization. In this case, the function (in the psychological sense) of an explanation is to help the subjects to reflect on their schemas, to detach them from their original sets of examples, which are now having a restrictive effect, and to modify them appropriately. The extension of index notation to zero and negative and fractional numbers would offer an example of this, if the new idea were presented in advance of the communications necessary to make it understandable. This seems to me to be a perfectly suitable way of teaching. It is not desirable never to put before the learner anything which does not relate by easy stages to what is already known. This 'over-programming' offers no challenge and no variety. It is often valuable to look first at the problem—say, that of finding the instantaneous velocity of a body in free fall; next, to analyse the tasks involved, which here include deciding on a suitable meaning for 'instantaneous velocity'; and methodically to develop, by the processes described, the new concepts (such as that of a limit) required to solve the problem.

## MAKING POSSIBLE REFLECTIVE ACTIVITY

This involves becoming aware of one's own concepts and schemas, perceiving their relationships and structure, then manipulating these in various ways. These three functions of the reflective system are represented in the diagram by the rectangles labelled 'receptors,' 'intervening mental processes' and 'effectors.' In the present context these intervening processes are cognitive, and make possible the overall activity which we call reflective intelligence. These are not, however, the only intervening processes which may occur: another variety will be discussed in Chapter 7.

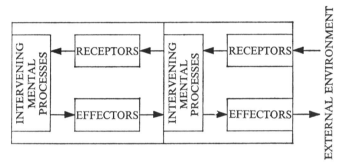

The process of becoming aware of one's concepts for the first time seems to be quite a difficult one. As was mentioned in Chapter 2, the overall development of this ability extends over a number of the years of childhood. But even in persons with a highly developed reflective ability, it is still a struggle to make newly formed, or forming, ideas conscious.

Making an idea conscious seems to be closely connected with associating it with a symbol. Just why this should be so is not yet known. Concepts are elusive and inaccessible objects, even to their possessors, and it may be that symbols (which are themselves primary concepts) are the most abstract kind of concept of which we can be clearly aware. Certainly more knowledge of the process would greatly increase the power of our thinking. Once the association has been formed, the symbol seems to act as a combined label and handle, whereby we can select (from our memory store) and manipulate our concepts at will. *It is largely by the use of symbols that we achieve voluntary control over our thoughts.*

Verbal thinking (which can be extended to include algebraic and any other pronounceable symbols) is internalized speech, as may be confirmed by watching the transitional stages in children. The use of pronounceable symbols for thinking is closely related to communication; one might describe it as communication with oneself. So becoming conscious of one's thoughts seems to be a short-circuiting of the process of hearing oneself tell them to someone else. This view is supported by the common observation that actually doing so to a patient listener (thinking aloud) is nearly always helpful when one is working on a problem. Visual thinking is a much more individual matter, and the relation between these two kinds of imagery will be discussed further in the next chapter.

## HELPING TO SHOW STRUCTURE

This function of symbols is related to that of the preceding section, since one of the aims of reflection is to become aware of how one's ideas are related, and to integrate them further. But the span of immediate memory is small: that is, the amount of information which one can keep in consciousness at a time is very limited. Moreover, the more difficult the topic, the more one needs to concentrate one's attention on one thing at a time. But one also needs to be able to refer quickly and easily to previous work. So one records one's thoughts on paper as one progresses. This is a more permanent form of the 'thinking aloud' discussed in the previous section, which reduces the cognitive strain of keeping the whole of the relevant information accessible.

Another way of reducing the cognitive strain, and a powerful one, is the brevity of mathematical notation. Compare:

| | |
|---|---|
| $(x + a)^2 = x^2 + 2ax + a^2$ | The square of the sum of two numbers is equal to the sum of their squares plus twice their product. |
| $Df$ | the derivative of (the function) $f$. |
| $D^{-1}f$ | the antiderivative of $f$. |
| 2751 | two thousand seven hundred and fifty one. |
| $\delta > O; EN:$ <br> $n \geqslant N, \lvert x_n - x \rvert < \delta$ | Given any positive number $\delta$, there is a number $N$ such that for all values of $n$ which are equal to or greater than $N$, the difference between the $n$th term of the sequence $x_1, x_2, x_3 \ldots$ and $x$ is less than $\delta$. |

But there is more to it than this. A good or bad symbol system can be a great help, or a severe hindrance, in evoking and manipulating the right concepts in the right relationships. Here is one example. First, do a simple multiplication, say, $34 \times 7$, in the familiar place value notation. Now, repeat (if you can) in Roman numerals: XXXIV multiplied by VII.

Another example:

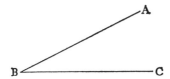

A conventional way to name this angle is $\angle ABC$. This suggests that the vertex of the angle is at $A$, which it is not. If there is no ambiguity, we refer to $\angle B$, not $\angle A$. The alternative usage $A\hat{B}C$ is better from this point of view, though less popular with printers. But let us now think about the ideas which we want to symbolize.

An angle is determined by two directions through a point. Each direction can be represented by a ray through that point, so an angle is represented by two rays, $\vec{BA}$ and $\vec{BC}$. If we are concerned with a *turn* from one direction to another, represented by a signed angle (the convention is to take an anticlockwise turn as positive), this will be represented by an ordered pair of rays $(\vec{BA}, \vec{BC})$, which can be condensed to $B_A^C$ and, where no ambiguity would result, to $B$.

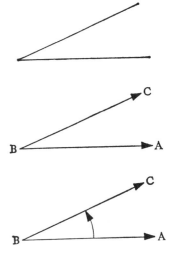

Addition and subtraction of angles at a point can be reduced to simple algebra once the 'rule of cancellation' has been seen and memorized.

$$B_A^C + B_C^D = B_A^D$$

Subtraction of an angle is equivalent to adding its inverse, that is, the angle corresponding to a turn in the opposite direction.

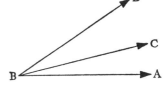

So
$$B_A^D - B_C^D = B_A^D + B_D^C$$
$$= B_A^C$$

What happens if we try to subtract a larger angle from a smaller? Let us try.

Algebraically,    $B_C^D - B_A^D = B_C^D + B_D^A$
$$= B_C^A$$

To what does this correspond in the figure?

This angle

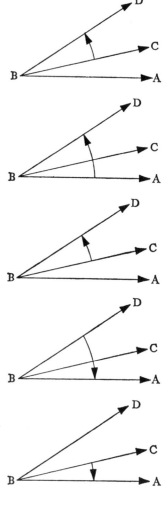

minus this angle

is equal to this angle

plus this angle

which is equal to this angle: which makes sense
in terms of successive turns.

Reverting now to the first equation, $B_A^C + B_C^D = B_A^D$
and comparing it with $\overrightarrow{AC} + \overrightarrow{CD} = \overrightarrow{AD}$, the reader who is familiar with the
addition of free vectors will recognize that the whole of the above algebra of
(signed) angles is isomorphic with that for the addition and subtraction of free
vectors, letters being read in the generally accepted positive directions of upwards and left to right.

There is no hope of replacing the long-established notation for angles by the above—the only hope of establishing a good notation is at the outset. Readers may not, in any case, be convinced by these arguments in favour of it. But it is hoped that they will at least agree that the choice of a notation should be made with care, and with some attention to how effectively it represents the ideas which (as this chapter has been devoted to showing) are so dependent on it.

The last example in this section was brought to my attention by Mr J. S. Friis. It requires an elementary knowledge of modular arithmetic and groups.

It is easily verified that $\{0, 1, 2, 3\}$ is a group under $\oplus$ mod 4, and so is $\{1, 2, 4, 3\}$ under $\otimes$ mod 5. The combination tables are, respectively,

| $\oplus$ mod 4 | 0 | 1 | 2 | 3 |   | $\otimes$ mod 5 | 1 | 2 | 3 | 4 |
|---|---|---|---|---|---|---|---|---|---|---|
| 0 | 0 | 1 | 2 | 3 |   | 1 | 1 | 2 | 4 | 3 |
| 1 | 1 | 2 | 3 | 0 |   | 2 | 2 | 4 | 3 | 1 |
| 2 | 2 | 3 | 0 | 1 |   | 4 | 4 | 3 | 1 | 2 |
| 3 | 3 | 0 | 1 | 2 |   | 3 | 3 | 1 | 2 | 4 |

It can be seen from these tables that these two groups are isomorphic, both being cyclic groups of order 4.

But if the right-hand table is rewritten in powers of 2, the isomorphism is revealed rather nicely.

| $\otimes$ mod 5 | $2^0$ | $2^1$ | $2^2$ | $2^3$ |
|---|---|---|---|---|
| $2^0$ | $2^0$ | $2^1$ | $2^2$ | $2^3$ |
| $2^1$ | $2^1$ | $2^2$ | $2^3$ | $2^0$ |
| $2^2$ | $2^2$ | $2^3$ | $2^0$ | $2^1$ |
| $2^3$ | $2^3$ | $2^0$ | $2^1$ | $2^2$ |

The reader will also observe how smoothly index notation generalizes to the new context. This, surely, is another criterion for a good notation.

## MAKING ROUTINE MANIPULATIONS AUTOMATIC

Thinking is hard work. Once we have understood a mathematical process, it is a great advantage if we can run through it on subsequent occasions without having to repeat every time (even though with greater fluency) the conceptual activities involved. If we are to make progress in mathematics it is, indeed, essential that the elementary processes become automatic, thus freeing our attention to concentrate on the new ideas which are being learnt—which, in their turn, must also become automatic. At any level, we can also distinguish between routine manipulations and problem-solving activity; and unless the former can be done with minimal attention, it is not possible to concentrate successfully on the difficulties. The same is true of any skill. To be a good driver, one must be able to change gear without thinking. Violinists cannot give themselves to the interpretation of the music until their technique is effortless.

In mathematics, this is done by detaching the symbols from their concepts and manipulating them according to well-formed habits without attention to their meaning. This automatic performance of routine tasks must be clearly distinguished from the mechanical manipulation of meaningless symbols, which is not mathematics.[5] Machines do not know what they are doing. Mathematicians, working automatically, can at any time they wish pause and reattach meanings to the symbols; and they must be able to pass easily from one form of activity to the other, according to the requirements of the task.

The economy of effort involved is striking. First, we learn to manipulate concepts instead of real objects; then, having labelled the concepts, we manipulate the labels instead. (And if the manipulations can be reduced to a mechanical process, we can even program a computer to do them for us.) Finally, perhaps, we reverse the process by re-attaching the concepts to the symbols and then re-embodying the concepts in the real actions with real objects from which they were first abstracted. So we calculate, say, the stresses involved, and design (mechanical) structures to withstand these stresses, for an aeroplane to fly at twice the speed of sound—before it leaves the ground and even before the first plates are riveted together. The power of mathematics is immense, and at all stages symbols make a major contribution to this power. But without the ability of mathematicians to invest them with meaning, they are useless.

## RECOVERING INFORMATION AND UNDERSTANDING

This function of symbols is somewhat like those discussed in section (ii), the recording of knowledge, and section (vi), in which symbols were described as a combined label and handle for identifying and manipulating concepts. Here we are concerned with using them for bringing concepts and schemas back into availability from one's long-term memory store. Even concepts in current use are elusive objects; those which have not been employed for some time may be quite inaccessible without, so to speak, some kind of handle whereby to pull them back.

Try this experiment. Ask a man what shape a reflector must be to give a parallel beam. Does he reply 'It is a surface formed by revolving about an axis of symmetry a curve having the property that for some point $S$ on the axis, and all points $P$ on the curve, a ray $SP$ and the line through $P$ parallel to the axis are

---

[5]We need to separate the two meanings, and the use of the two words automatic and mechanical seems a convenient way to do so.

equally inclined to the tangent at $P$ to the curve. This curve is called a parabola.' Or does he immediately reply 'a parabola,' afterwards explaining the properties just described? In other words, does he first recall the conceptual structure, or does he first recall its label?

Another example, for those with some knowledge of quadratic equations. Has the following equation real roots? $3x^2 - 4x + 2 = 0$. In the majority of cases, either the word 'discriminant' or the symbols '$b^2 - 4ac$' will first come to mind. Afterwards, the method based on this will be recalled.

One more example. Have you ever met, say, old schoolfellows or colleagues and not recognized them; but as soon as they said 'I'm . . .,' you not only recognized but remembered much else about them?

This process of recovery of information by the help of symbols is exemplified by all mnemonics. One example will suffice.

| sin | all |
|-----|-----|
| tan | cos |

This is a well-known device for remembering the signs of the trigonometrical ratios of angles from 0° to 360°. The diagram gives the positive ratios.

The difference between a mnemonic and a formula is that the latter embodies the structure of what is to be recalled. From a formula, therefore, understanding can be reconstructed, even if it does not immediately follow the recall of the symbol. By hearing or seeing the words 'Ohm's law,' the formula[6] $E/I = R$ will, for many persons, be evoked—that is, it will come into present consciousness from the long-term memory store. From further consideration of the formula it is easy to reconstruct its meaning: that for a given circuit the ratio of electromotive force to current is constant—double the volts and the amps will also be doubled.

This order of recall—symbols first, then meaning—is not, however, invariable. When recalling a conversation, or something read, most people reproduce the meaning but express it in their own words. Similarly, teachers sometimes say 'Don't memorize this result;[7] it is easy to reconstruct it when you want it.' (Example: the equation of the tangent to some particular curve.) It is also certain that the initial process of memorizing is very much easier for symbols with meaning than for meaningless material.

In mathematics, what we *store* is a combination of conceptual structures with associated symbols, and the former would therefore seem to be important for the *retention* of the whole. The question is, which part of the symbols-and-concepts combination is easiest to 'catch hold of' when we are trying to *recall* material from this store into consciousness? And although there seems to be some evi-

---

[6]Or one of its equivalent forms.

[7]Meaning, of course, 'the symbols representing this result.'

dence that it is by symbols that recall is most easily achieved, this view being consistent with the other functions of symbols, the situation is not entirely clear. Further research is needed.

## CREATIVE MENTAL ACTIVITY

In one sense, since all new learning in mathematics by the method of concept-building consists of the formation by individuals of new ideas in their own minds, it is creative from their point of view. This is why, learnt in this way, mathematics is an exciting pursuit. But the description is used more particularly for the creation of ideas which no one else has had before—for opening up new paths, rather than retracing existing ones, though the latter are new to the learner following them for the first time. The former is an unreliable process and may take years. Once the new insight is achieved, it can be communicated in the ways already discussed to all others whose schemas are far enough developed in the right direction to be able to assimilate it.

Ghiselin (1952, 1955), in his classic work *The Creative Process* has collected together reports from originators in many fields—musicians, writers, scientists, mathematicians. From these, what emerges perhaps most clearly is that this process will not perform to command. The central part of the activity is both unconscious and involuntary.

There is, however, a fair degree of agreement that a necessary preliminary is a period of intense concentration on the problem. Following this, there is usually a period when the problem is laid aside, so far as the conscious mind is concerned, a period of relaxation, other mental or bodily activity or sleep. Apparently, during this period, unconscious mental activity concerned with the problem continues, for suddenly an insight relating to the problem—perhaps a complete solution—erupts into consciousness, at a time when no deliberate work on the problem is in progress. This insight is accompanied by a strong feeling of pleasure, and, interestingly, an urge to communicate.

What part do symbolic processes have in this creative activity?

Since the central stage, in which existing ideas suddenly fit together in a new way to produce an altogether new idea, is unconscious and involuntary, it is impossible to say whether, or to what extent, symbols play an essential or a contributory part here. In the preceding and the succeeding stages, however, their function is essential.

The first stage is that of intense, often prolonged, concentration on the problem, in which all the relevant ideas are brought together and considered from many aspects and in many different combinations and relations to each other. (Not all, for at this stage the insightful combination is not produced.) During this period of reflection, symbols play an essential part, for it is by their use that we achieve voluntary control over our thoughts. It may well be that it is at this stage that the contributory concepts are raised to a high enough degree of activity for the ensuing synthesis at the unconscious level.

When the insight does occur, it may well attach itself spontaneously to suitable symbols, for this seems to be closely associated with the process of making conscious. But this is likely to be incomplete, and the symbolization has to be continued deliberately, to make possible the communication and recording of the results of the creative process. This formulation and recording, with which is closely associated the process of making fully conscious, is often described as a painful struggle.

Unfortunately, too, not all the ideas which arrive in this way fulfil their early promise. After the insight must come verification. In science, this means the testing of the idea by experiment. In mathematics, it means logical analysis, testing for internal consistency and consistency with accepted knowledge. This is again a reflective process, for which symbols are essential. They may also, if chosen with care (which unfortunately is not always the case), contribute importantly to revealing the new structure.

*

The length of this chapter (it is the longest so far) is a measure of the importance of symbols in the learning and use of mathematics. There are two ways in which a relevant concept may be evoked: by encountering an example, which evokes it intuitively and involuntarily; and by the use of an associated symbol, which makes possible voluntary control, communication and the recording of knowledge. English and mathematics have both been described by Bruner as 'a calculus of thought,' and it is their symbol-systems which make them so. Without an appropriate language, much of the potential of human intelligence remains unrealized.

# 6

# Different Kinds of Imagery

As long ago as the 1880s, Galton (1822–1911) found that people differed greatly in their mental imagery. Some, like himself, had strong visual imagery; others had none at all, and thought mainly in words. This is as true today as it was then, and there are also individuals who have available both, though often with a preference for one or the other modality. (It is not, however, always easy to decide what kind of images people use, or indeed whether they have any at all.) In this chapter we shall be considering the two kinds of symbol, visual and verbal, which are used in mathematics, both in mental imagery and for all the other purposes served by symbols.

## VISUAL SYMBOLS AND VERBAL SYMBOLS

First, these terms need clarifying, for as soon as words are written down they become things to be seen not heard. Nevertheless, words begin as auditory symbols, and their primary mode of communication is by word-of-mouth not word-on-paper. A reader usually turns written words into sub-vocal speech (though teachers of rapid reading point out that this is time-wasting). So by 'verbal' we shall mean both the spoken and the written word.

Visual symbols are clearly exemplified by diagrams of all kinds, particularly geometrical figures. But into which category should we put algebraic symbols like these?

$$\int_a^b \sin x \, dx$$

$$\{x : x^2 \geq 0\}$$

Basically these are a verbal shorthand. They can be read aloud, or communicated without ever taking a visual form. This first is read as 'The integral from $a$ to $b$ of sine $x$ $dx$' (or, '. . . with respect to $x$'); and the second as 'The set of all values of $x$ such that $x^2$ is greater than or equal to zero.' The advantages of the algebraic notation are, first, those of any shorthand—a saving of time and trouble. But this brevity also adds greatly to its clarity and power, since the individual ideas for which they stand are evoked in a much shorter space of time, favouring apprehension of the structure as a whole. There may be less tendency to read them sub-vocally, and there are certain visual aspects which will be mentioned later. But as further discussion will show, algebraic symbols have much more in common with verbal symbols than they have with diagrams and geometrical figures, and for the present they will be classed with the former. A supporting argument is the way that verbal and algebraic symbols are mixed: for example, 'If $p$ is a prime number, then $p \mid ab \Rightarrow p \mid a$ or $\mid b$.' ('If $p$ is a prime number, then $p$ divides $ab$ implies that $p$ divides $a$ or $p$ divides $b$.')

Both visual and verbal symbols are used in mathematics, together and apart. Thus we find diagrams with verbal explanations and, say, trigonometrical calculations; we find curves together with their equations; but we also find page after page of algebra with no kind of figure or diagram. Indeed, a recent and highly thought-of book on geometry also contains not a single figure! It looks as if verbal (including algebraic) symbols are indispensable, but visual symbols are not.

Even if they are not indispensible, however, there is no doubt that visual symbols are often very useful and may be a great deal more understandable than a verbal-algebraic representation of the same ideas. One sometimes also has the impression that the avoidance of diagrams is a demonstration, perhaps unconscious, that the writer needs no such props to his thinking—an academic 'Look, no hands!'

A reasonable working hypothesis is that the functions which these two kinds of symbol perform are different, perhaps complementary. If this is so, we want to know what these functions are, with a view to using and combining them to best advantage. For, let it be repeated, the part played in mathematics by symbols is crucial (see again the list on page 46). So any improvement in our knowledge of how to choose and use symbols, and devise new ones, would have great potential value.

Visual symbols seem to be more basic, at least in their primitive form of representations of actual objects. As Piaget has shown, even our perception of an object involves a kind of concept, though quite a low-order one. When we see any object from a particular viewpoint on a particular occasion, this experience evokes a memory of all our earlier experiences of seeing this object—not separately but as an abstraction of something common to this class of experiences. This is experienced as 'recognition,' and we endow the object, in the present experience, with various other properties which derive not from the incoming sense-data but from the object-concept which is evoked. So a visual image, or a pictorial representation, of an object may fairly be described as a symbol, though the associated concept (that of the object) is of lower order than those used in mathematics.

By leaving out quite a lot of the visual properties of an object we can abstract at a higher level, while still representing the resulting concept visually. Maps, circuit diagrams and engineering drawings are all examples in which the most important properties of an object can be much better represented by visual than by verbal symbols.

For a mathematical example, consider this diagram, which represents a tall block of flats on level ground. For present purposes we are only interested in its height and shape.

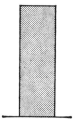

Next we have represented a surveyor's observation of the angle of elevation of the top of the building, taken at a distance of 100 metres from the base. It is interesting here to note that the surveyor himself and the direction of his observation are both represented by spatial symbols (points and lines), while the measurements and the unknown height are represented by verbal-algebraic symbols.

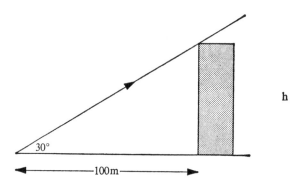

Already we need both, and as soon as the calculation begins, we go over completely to the latter.

$$h = 100 \tan 32°$$

Nevertheless, the diagram is of great help initially in representing the overall structure of the problem. It gives the context from which the particular calculation needed is abstracted.

Although they are more basic, visual images are much more difficult to communicate than auditory ones. For the latter, all we have to do is to turn our vocal thinking into speaking aloud. But to communicate our visual thoughts, we have to draw or paint or make a film. Nowadays, computer graphics make this easier, but the process is still much harder than speaking. This gives verbal communication a great advantage over visual. Moreover, the bringing into consciousness of an idea is closely linked with the use of an associated symbol. Since we hear our own speech at the same time as does the listener, the same ideas are evoked nearly simultaneously into the consciousness of both—provided, of course, that the symbols used have approximately the same meanings for both. So when speaking our thoughts to another, we are also communicating them to ourselves.

There is also some evidence that audible speech brings ideas into consciousness more clearly and fully than does sub-vocal speech. When working more difficult problems in arithmetic, children often drop back into whispering their thoughts, and in his voyage single-handed around the world in *Gypsy Moth IV*, Sir Francis Chichester found, when working the boat in difficult conditions and when very tired, that it helped to tell himself aloud what was to be done. This would explain the common experience that after simply stating a problem (academic or otherwise) aloud, even to a hearer who makes no contribution other than to listen, we sometimes find a solution.

When a discussion takes place, we get this subjective effect on both sides, together with the interaction of ideas which is the more conscious purpose of those taking part. The resulting progress of thought can be considerable. Because it is so easy to transmit our verbal symbols, and so much harder to transmit our visual symbols—we have built-in physical apparatus for the former but not the latter—the double advantage described above is attached, in the experience of most of us, much more strongly to verbal symbols.

## Socialized Thinking

It follows from this that our verbal thinking is likely to be more socialized, since it is to a greater extent the end product not only of our individual thinking but of that of others, and of interaction between the two. To see things, literally, from someone else's point of view, we would have to go and stand where they were, or receive from them a drawing or a photograph, whereas they can talk about what they see without our making a move, and we can both hear the same sounds while standing in different places and looking in different directions. Vision is individual, hearing is collective, at the concrete as well as at the symbolic level. And it is interesting to notice that when we do wish to emphasize individual rather than collective aspects of a set of ideas, we talk about a 'point of view.' Even 'aspect' is a visual metaphor. So a contrast is beginning to emerge between the two kinds of symbols along the following lines.

Visual: harder to communicate, more individual.

Verbal: easier to communicate, more collective.

A human being is a social animal, and the advantages of communication are so great that the predominance, noted earlier, of verbal thinking might well be explicable on these grounds alone. But the advantage of communicability is an accidental one (we have built-in loudspeakers but not built-in picture projectors) and not intrinsic to the nature of the symbols themselves. Indeed, it is sometimes said that 'one picture is worth a thousand words.' If this is so, then instead of writing this book (about 90,000 words), the author would have spent his time better in making ninety pictures. With modern techniques of reproduction, this would have presented no difficulty of publication. Moreover, the written word loses the particular advantages of simultaneity for speaker and hearer which the spoken word has, and the interaction between them. So, is writing books and reading them, rather than drawing them and looking at the pictures, simply a habit taken over from the habit of conversation and discussion? Or are there also intrinsic advantages in the verbal-algebraic kind of symbol?

## Visual Symbols in Geometry

Geometry suggests itself as a profitable context in which to investigate this question, since this is one of the areas of mathematics in which diagrams seem to have particular importance. We must note at once that the symbols involved are more abstract than a visual representation of an object. Even a life-size colour photograph of an object shows only a single aspect, and, to the extent that it evokes the concept of the object-as-totally-experienced, it could be described as a symbol for the object. Other representations abstract further, usually showing shape rather than colour, texture, size. Another degree of abstraction is found in drawings which represent not a particular object but a class of objects. Even photographs may serve this purpose—one which advertises a new model of car is intended to persuade us to buy not that particular car but one of a particular class of cars. We attribute to it every property common to all members of that set— acceleration, speed, comfort, etc.—but no particular quality, such as engine number, colour. The photograph is just as much a symbol for a *variable,* in the strict mathematical sense, as, say, the words 'Jaguar XJ 420.'

A major difference between the two kinds of symbol, photograph and words, is that one looks like a typical object of the set which it represents, whereas the other does not sound like it. So this visual symbol, at any rate, has a closer link with the concept than has the corresponding verbal symbol. The same is true of geometrical symbols. This is a geometric symbol:

This is the corresponding verbal symbol: a circle.

The resemblance of the geometric symbol to its concept has both advantages and disadvantages. An advantage is that it evokes well the properties of the concept. This is especially so when we represent visually several concepts together. The diagram then brings into awareness the relationships between these concepts far more clearly than does a verbal representation of the same concepts.

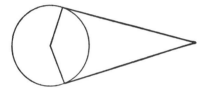

A circle, two tangents to it from a point outside the circle, and the radii through the points of contact of the tangents.

A disadvantage of the visual symbol is that it has to be drawn to be communicated—but pencil and paper, chalkboard and chalk, are easy enough to use. We also have to remember that it represents not a particular circle, tangent, etc., but variables—*a* circle, not *the* circle of given radius and centre which we see, etc. The words remind us explicitly of this. Since the diagram cannot but show a particular circle, etc., we have to remember to ignore its particular qualities and work with the general ones which it symbolizes. Because it is a stage more concrete, we must do some of the abstracting ourselves.

In the present example, however, these two minor disadvantages are quite outweighed by the conciseness and clarity of the visual symbols. Nevertheless we find that while most geometrical communications begin with a diagram, they very soon change over to verbal-algebraic symbols, together with additional geometrical ones such as $\widehat{AOB}$, ⊢, ‖ . And the visual element is sometimes abolished altogether. In the study of vectors, directed line segments are replaced by ordered pairs, triplets or n-tuples of numbers; and one of the directions in which geometry currently seems to be moving is that of an algebraically manipulated axiom system. Why does not this, one of the most visual branches of mathematics in its early stages, remain so?

## Visually Presented Arguments

The following examples suggest that we might, with advantage, stay in the visual mode more than we do at present. With a few simple conventions, the diagrams below convey all that the verbal statements do, more clearly and vividly.

The tangents to a circle from a point outside it are equal in length.

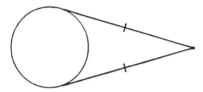

Note that the diagram also shows which parts of the tangents we mean, which in the verbal statement is left implicit and could only be made explicit by so many extra words that the meaning would then be less clear than before.

The exterior angle of a triangle is equal to the sum of the interior opposite angles.

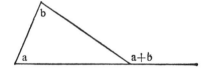

This is the usual statement. We should really say 'the size of the exterior angle', since an object and the size of an object are different ideas. This shows more clearly in the diagram, where the angles are represented by pairs of lines and their sizes by letters. And who would know which angles we meant by 'exterior' and 'interior opposite' without a diagram? Here the verbal statement is much inferior to the visual.

We can also show a theorem and its converse. The angle in a semi-circle is a right angle.

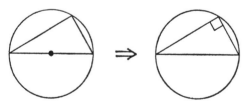

Here $\Rightarrow$ means 'implies.' The left-hand figure shows the data, using the convention that a dot drawn approximately at the centre of a circle does in fact represent the centre. The right-hand figure represents the conclusion derived by this theorem from the data.

The converse of this theorem is also true. If a chord of a circle subtends a right angle at the circumference, that chord is a diameter.

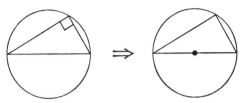

By using the sign ⇔ for a two-way implication, we can represent simultaneously both the theorem and its converse. The angle in a semi-circle is a right angle. Also, if a chord of a circle subtends a right angle at the circumference, that chord is a diameter.

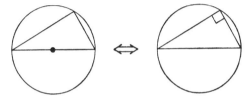

So far, the visual statements are much clearer and briefer. Difficulties begin to arise when we want to do two more things—give a logical proof and direct attention to particular parts of the diagram. The first of these often necessitates the second.

The above theorem is a particular case of the following. The (size of the) angle at the centre of a circle is twice the (size of the) angle at the circumference subtended by the same chord or arc.

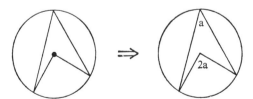

The proof of the earlier theorem consists in pointing out that we may consider

this straight line 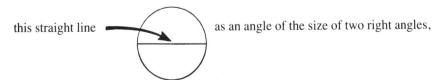 as an angle of the size of two right angles,

having its vertex here       at the centre of the circle.

The theorem given last tells us that this angle

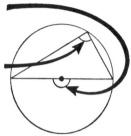

is twice the size of this angle.

But the size of this angle is two
right angles,

so the size of this angle is one
right angle.

This is still clear, but much clumsier. In a face-to-face situation the same diagram would be used throughout, and the speaker would point to the parts of the diagram being referred to at the appropriate moments. The stumbling-block is the translation of an act of pointing into a diagram. Once we have drawn an arrow, we cannot erase it in a way which corresponds to the withdrawal of one's hand; we have to re-draw the diagram. And the arrows also clutter the diagram, because they are too like part of it. Different colours would help.

Another use of words has been to suggest new classifications to the reader: for example, that a straight line may be considered as a particular kind of angle. This can also be shown visually. It takes more space, but is more vivid. There is a

certain resemblance to a strip-cartoon, and if one has the resources and ability to translate this into computer graphics, the visual presentation can retain all its advantages. What would be the stages of such an animation? The following is one possibility. Note that the first figures represent the data.

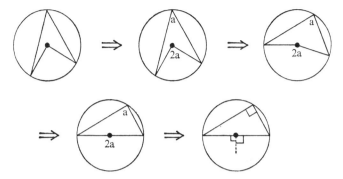

For comparison, here is a conventional proof of the same theorem.

*Data*      $AOB$ is a diameter of a circle, centre $O$.
            $P$ is any point on the circumference.
*To prove*  $APB = 1$ rt $\angle$
*Proof*     $AOB = 2\ APB$ ($\angle$ at centre $=$ twice $\angle$ at circumference)
But         $AOB = 2$ rt $\angle$ s, since $AOB$ is a straight line.
        $\therefore\ APB = 1$ rt $\angle$

Q.E.D.

Here we use letters as a substitute for pointing. When the letters are found in the (verbal-algebraic) proof, we then have to find these letters in the diagram, and this tells us where to look. This is neater than the long arrows used on page 97 and saves re-drawing the diagram. Which is the easier to follow, the reader must judge for himself. This too could with great advantage be translated into computer graphics.

How does the 'purely visual' approach cope with more complex proofs? Space must limit us to one further example—a proof of this more general theorem already referred to.

*Theorem*

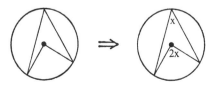

*Proof*

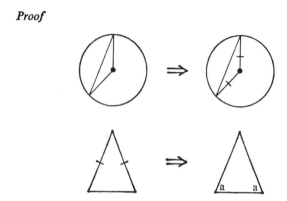

Is this clearer than a verbal-algebraic proof (for which, see any traditional school geometry text), or is it another case of 'look, no hands'—this time, no words? Since individuals differ in their preferences for visual or verbal-algebraic symbolism, there may be no general answer to this question. At present the latter system has achieved dominance. The chief purpose of the foregoing has been to question this *fait accompli* and examine the particular contribution of visual symbolism.

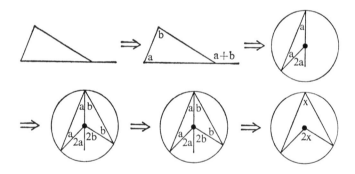

## THE TWO SYSTEMS IN CONJUNCTION

Historically, one of the happiest marriages of the two systems is that due to Descartes (1569–1650). Any point in the plane of the paper is specified by its distances from two (usually perpendicular) lines, that is, by two numbers, written as an ordered pair. These *coordinates,* as they are called, may be positive or negative. A variable point corresponds to a pair of numerical variables; and a set of points with a given characteristic property, for example, that their distance

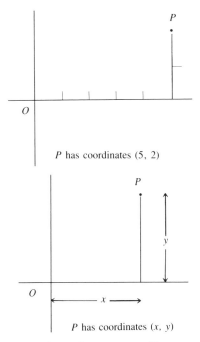

$P$ has coordinates (5, 2)

$P$ has coordinates $(x, y)$

from the origin is always equal to $r$, is represented by an equation satisfied by all the pairs of coordinates $(x, y)$. By these means curves which are difficult to draw

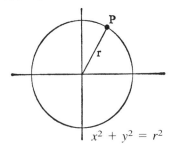

$x^2 + y^2 = r^2$

accurately can be represented algebraically: for example, an ellipse, which is the shape of a planet's orbit round the sun; a parabola, which is the shape a reflector

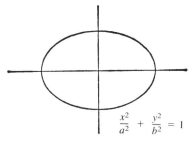

$$\frac{x^2}{a^2} + \frac{y^2}{b^2} = 1$$

must be to give a parallel beam (as for a car headlight) or to concentrate distant rays to a point (as for a radio telescope). Both general and metrical properties can

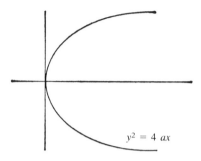

$$y^2 = 4\,ax$$

be dealt with in this way: general properties, by using general relations between variable coordinates, and metrical properties, by giving particular numerical values to these variables. What this algebraical treatment of geometry adds is great power of manipulation and accuracy far beyond what is available by accurate drawing to scale and measurement of the drawing. But we still need the drawing to show what the set of points looks like as a whole. It is, for example, not obvious from the equations that the curve represented by $y^2 = 4ax$ disappears into the distance in two directions, or that the curve represented by

$$\frac{x^2}{a^2} - \frac{y^2}{b^2} = 1$$

joins itself again, or that a simple change of sign in the latter will give us something looking completely different.

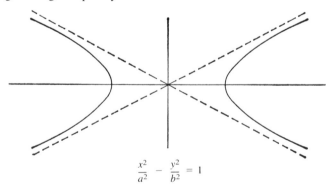

$$\frac{x^2}{a^2} - \frac{y^2}{b^2} = 1$$

That neither kind of representation is superior in all ways is suggested by the fact that we often use the method in reverse. Instead of starting with a known curve (all the above were known to Greek geometers about eighteen centuries before Descartes) and representing it algebraically, we may start with an algebraic concept, that of a function, and represent it graphically.

The idea of a mathematical function is one of great generality. Broadly speaking, functions tell us how the objects in one set correspond to those in another: for example, how the distance travelled by an object may be found if we know the time; how the current through a given circuit may be determined if we know the voltage. Functions may be represented in a variety of ways, including equations and graphs.

For finding individual correspondences, an equation is very convenient. For example, if $d$ metres is the distance travelled by a body in free fall under gravity (neglecting air resistance) and $t$ seconds the time it has been falling, then $d = 4 \cdot 9\, t^2$. So the distance fallen after one second is $4 \cdot 9 \times 1$ metres, after two seconds it is $4 \cdot 9 \times 4$ metres, and so on. By taking $(t, d)$ as Cartesian coordinates, we can show graphically the function as a whole.

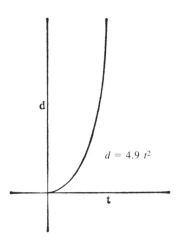

$$d = 4.9\, t^2$$

## THE TWO SYSTEMS COMPARED

Tentatively, we may now attempt a summary of the contrasting, and largely complementary, properties of the two kinds of symbol.

| *Visual* | *Verbal-algebraic* |
|---|---|
| Abstracts spatial properties, such as shape, position | Abstracts properties which are independent of spatial configuration, such as number |
| Harder to communicate | Easier to communicate |
| May represent more individual thinking | May represent more socialized thinking |
| Integrative, showing structure | Analytic, showing detail |
| Simultaneous | Sequential |
| Intuitive | Logical |

The communicable, socialized properties of the verbal-algebraic system have doubtless contributed to its predominance over the visual system. Yet whenever we want to represent also the overall structure of some topic, argument or situation, visual symbolism returns, as in organization charts (from firms to football teams), flow diagrams and family trees. The value of visual symbolism is also shown by the way in which it superimposes itself on the verbal-algebraic, in the form of spatial arrangement of written symbols. Auditory symbols are inevitably sequential in time. When written down, they are present simultaneously, the sequential arrangement being restored by scanning them in a conventionally agreed order. But this order may be departed from whenever we like. We may look quickly at the beginning and conclusion of an argument before examining details. We may recapitulate whenever we wish, and this becomes necessary more often as the argument becomes more involved. In other words, a verbal-algebraic exposition, once written down, shows the overall structure in addition to the logical-sequential implications within the structure, and it may be scanned in other ways besides the conventional left to right, top to bottom, order.

Spatial symbolism finds its way into every detail of the verbal-algebraic system.

| | |
|---|---|
| The position of a digit helps to show what number it represents. | 2      7     3 <br> 2 hundreds, 7 tens, 3 units. |

Position shows which number is subtracted from which,

$$9 - 5$$

or divided by which.

$$\frac{16}{4}$$

| | | | | |
|---|---|---|---|---|
| Position shows correspondence between two sets, as in this proportion. | 1 <br> 4 | 2 <br> 8 | 3 <br> 12 | 4 <br> 16 | 5 <br> 20 |

Its spatial arrangement is an essential property of a matrix.

$$\begin{pmatrix} a_1 & a_2 & a_3 & a_4 \\ b_1 & b_2 & b_3 & b_4 \\ c_1 & c_2 & c_3 & c_4 \end{pmatrix}$$

Many other examples could be given.

Before concluding this chapter, it will be interesting to return briefly to the individual differences in imagery noticed by Galton and mentioned at the beginning. If we are right in thinking that visual imagery is that most favourable to the integration of ideas, and if it is not accidental that when we first become aware of how ideas relate to each other, we refer to the experience as insight, not as in-hearing, then we might reasonably hypothesize that persons who have been noteworthy for their contributions to mathematical and scientific understanding will be found to use visual rather than auditory imagery.

We may as well begin the list with Galton himself, who tells us that his own visual imagery was clear but that he lacked verbal fluency. Einstein (1879–1955), in a letter to Hadamard, states that his preferred imagery is visual and motor, and that 'conventional words or other signs have to be sought for la-boriously, only in a secondary stage' (Hadamard, 1945). A famous non-mathe-matical example is that of Kekule (1829–96), whose conception of the ring stru:ture of the benzene molecule came to him in a dream where he saw a snake taking hold of its own tail. And the Nobel Prize winner Bragg (1890–1971), in a television programme honouring his eightieth birthday, stated that his new ideas came to him for the first time in the form of visual images.

This list is a partial and selective one, and we lack the comprehensive infor-mation about other famous mathematicians which would support or refute the hypothesis. An interesting discussion, along more general lines, of personality traits of mathematicians and others is to be found in Appendix 2 of Macfarlane Smith's *Spatial Ability;*[2] where much other material of interest in the present context may also be found.

Analysis, logical argument and socialized thinking are, rightly, much valued in mathematics, but we also need synthesis, insight and individual thinking. To some extent the former seem to be capable of being taught; the latter, at present, can only be sought. If we can discover more about the functions of the two kinds of symbol discussed in this chapter, and become ore skilled in choosin an using them, this might well help us to develop and relate these two comlementary aspects of our mathematical thinking.

## HEMISPHERIC SPECIALIZATION

Nine years after the first edition of this book was published, I attended a lecture by Glennon, subsequently published as a monograph entitled *Neuropsychology and the Instructional Psychology of Mathematics* (Glennon, 1980). Readers may judge for themselves the interest with which I heard, and subsequently read, the following passage. (To make even clearer the correspondences with the table on page 104, I have taken the liberty of interchanging the left- and right-hand columns in Glennon's table.)

In general, the left hemisphere processes verbal and analytic information. In general, the right hemisphere processes visuospatial and Gestalt (holistic) information. A summary of the findings from many research studies suggests the hemispheres perform these functions:

| Right Hemisphere Functions | Left Hemisphere Functions |
|---|---|
| Visuospatial (including gestural communication) | Verbal |
| Analogical, intuitive | Logical |
| Synthetic | Analytic |
| Gestalt, holistic | Linear |
| Simultaneous and multiple processing | Sequential |
| Structural similarity | Conceptual similarity |

Until this lecture, Glennon and I had not met, nor were we acquainted with each other's work.

# Interpersonal
# and Emotional Factors

This is primarily a book about learning mathematics with understanding, not about teaching it, though there are, of course, many implications for the latter. But most readers are likely to have the same attitude to the subject as they acquired at school, so an examination of these attitudes, and how they may have been acquired, is still relevant. For those with feelings of dislike, bafflement or despair towards mathematics, the aim of this chapter is to suggest that the fault was not theirs—indeed, that these responses may well have been the appropriate ones to the non-mathematics which they encountered. And those who remember their school mathematics with interest and pleasure will realize, if they did not before, how lucky they were. Chapters 2 and 3, in particular, have emphasized the particular dependence of the student of mathematics on good teaching, especially in the earlier stages, when foundation schemas, and also what may be long-lasting attitudes to the subject, are being formed.

Before contact with the learners (whatever age they are), the teacher of mathematics has two important tasks: first, to make a conceptual analysis of the material; second, to plan carefully ways in which the necessary schemas can be developed, with particular attention to stages at which restructuring of the learner's schemas will be needed. Then, when in direct contact with learners, the teacher is responsible for general direction or guidance of the work, for explanation and for correction of errors. The teacher also needs, to a varying extent, to create and maintain interest.

The before-contact tasks will usually be done by someone other than the face-to-face teacher. They are difficult and timeconsuming, and the teacher who is involved in the day-to-day work of teaching is seldom in a position to undertake them. Whoever does these—college lecturer or writer of mathematical texts—plays an essential part in the teaching process, but let us here for convenience restrict the term 'teacher' to the face-to-face teacher (or possibly correspondence-course tutor) who is in direct and continuing contact with the learner. In this chapter we shall be concerned with the personal interactions between teacher, in this sense, and learner, and the ways in which they may affect the learning of mathematics with understanding.

## WHAT CRITERION?

Mathematics has much in common with the natural sciences and less in common with languages and subjects like history, English literature. It differs from all these, however, in one important respect. In the natural sciences, the basic criterion for the validity of any statement or piece of work is experiment. Admittedly, not all the experiments will be done, or even witnessed, by the students. But in the main, if they are willing to accept in good faith that certain events result if certain conditions are set up, and particularly if they have some basic schemas based on their own experiments and observations, students of the natural sciences develop their knowledge in an interpersonal situation where the ultimate appeal is to facts and not to the authority of the teacher.

This is in marked contrast to some other subjects, for example, Latin, where the correctness of a piece of translation is decided on the authority of the teacher, or English, where again the final arbiter of the merits of an essay is the teacher (or examiner). In the former example, the teacher's opinion may be supported by the printed word, but this too is based on authority, not experiment. In the latter case, no appeal is available at all, except perhaps to another teacher—a second opinion, not an objective verification.

Where does mathematics stand in this? The question is important, because few people really like being told they are wrong, or otherwise diminished. But students are likely to accept this more readily if they can be given better evidence than 'because I say so,' whether expressed thus or more politely. So what is (or should be) the final criterion for the validity of a mathematical piece of work—solution of an equation, proof of a theorem or answer to a problem in mechanics?

Certainly in pure mathematics, the ultimate appeal is not to experiment. (By what laboratory experiment can one prove that the square root of $-1$ is not a real number?) Nor is it, or rather nor should it be, to the teacher's authority. The final criterion of any piece of mathematics is consistency. This may be within a particular piece of mathematics—any solution to an equation must satisfy the equation in its original form, and if students offer an incorrect solution, this is how any averagely good teacher will tell them to check. Or it may be within the larger mathematical system of which it forms part. Whether this consistency exists is a matter for agreement between one mathematician and another, and between teacher and learner. The interesting, and rather surprising, thing is the high degree of agreement which can be achieved on such a basis. What is more, the criterion is implicitly accepted as binding by teachers and students alike. If a teacher makes a mistake when working on the blackboard and a member of the class points it out, the teacher has no alternative but to correct it. Teachers are subject to the same rules as learners, and these are not the rules of an authoritarian hierarchy but of a shared structure of concepts. In mathematics perhaps more than any other subject the learning process depends on agreement, and this agreement rests on pure reason.

## Insults to the Intelligence

Learners have no need to accept anything which is not agreeable to their own intelligence—ideally they have a duty not to. And it is by the exercise of the teacher's intelligence, not by prestige, eloquence or tyranny, that the learners should be led to agree with their instructor. The teaching and learning of mathematics should thus be an interaction between intelligences, each respecting that of the other. Learners respect the greater knowledge of the teacher, and expect their own understanding to be enlarged.

Suppose now that what they encounter is not intelligent or intelligible material at all, but a series of meaningless rules: for example, that they must, to solve an equation, 'get all the $x$'s on one side and all the numbers on the other,' and that the way to do this is to 'take them over to the other side and change the sign.' (See page 86.) Instructions of this kind may fairly be described as a series of insults to the intelligence, for they purport to be based on reason but (usually) are not.

The term 'insult' is used here both in the everyday sense and in the medical sense of something injurious to an organism. Trying to understand something involves assimilating it to one's schemas. To the extent that what is being communicated is not intelligible, the receiver is trying to expand or restructure personal schemas to assimilate meaninglessness. To do this would be equivalent to destruction of these schemas—the mental equivalent of bodily injury.

Viewed in this light, one can begin to see why some learners acquire not just a lack of enthusiasm for mathematics but a positive revulsion. What is more, they are in these circumstances quite right in so doing, because one of their highest faculties, their developing intelligence, is being exposed to a harmful influence. That the teacher means no harm, but is only acting in ignorance, does not affect the situation at the receiving end. And it is, moreover, likely to be the more intelligent learners whose minds boggle at the unorganized collection of rules without reasons which often constitutes the teaching of so-called mathematics. They are aware that they cannot find meaning in what is presented to them, but unaware that the fault is not theirs. Either the matter, in the form presented to them, is not meaningful, or else they have not been given certain preliminary ideas necessary to the understanding of the new ones.

## RULES WITHOUT REASONS

This kind of teaching is as though someone learning to drive was told that whenever they wanted to come to rest they had to depress the clutch pedal as well as the brake, without ever having been told what was the function of the clutch pedal. 'Why?' they ask. 'If you don't, the engine will stop.' 'Why?' 'It just will.' The first reason is sound as far as it goes; but to answer the second 'Why?', two basic facts are needed. First, that an internal combustion engine will not, like an electric motor or a steam engine, start from rest under load. It has a minimum operational speed. Second, that to allow the engine to keep running independently of the gear box and road wheels, a gadget called a clutch is fitted which allows the engine to be connected to and disconnected from the gearbox at will.

'To divide by $\frac{2}{3}$, you multiply by $\frac{3}{2}$.' 'Why?' Readers are invited to search in their memories to find whether they have ever been given a good reason for this, or, alternatively, to seek an explanation from a school child of suitable age, to discover whether he or she has received any good reason.

The list of mathematical examples could be continued almost indefinitely, at both elementary and advanced levels. Some readers may remember learning to solve equations by some such method as the following, which is still in use: 'We use the rule that when we change the side we change the sign.'

To solve the equation we first get all the $x$'s on one side by taking the $x$ over and changing the sign.
Then we take $-3$ over to the other side and change the sign.
Simplify both sides.
Take the 5 across and change the sign.

$$6x - 3 = 7 + x$$
$$\therefore \ 6x - x - 3 \ = 7$$
$$\therefore \ 6x - x \qquad = 7 + 3$$
$$\therefore \qquad\qquad 5x = 10$$
$$\therefore \qquad\qquad x = 10 \div 5$$
$$\therefore \qquad\qquad x = 2$$

*Answer: x = 2*

If all that is wanted is to be able to solve equations of this kind quickly and efficiently, such a method is adequate. If, however, any importance is attached to understanding what one is doing, then it is not. And this understanding is not just a luxury which makes the task more pleasant: it is a necessity if one is to be able to adapt one's knowledge to new situations. The topology example given in Chapter 3 (page 30) was introduced to make just this point. In that example, the ideas which were necessary to convert the rule without reason into information which could be assimilated by the intelligence were few and simple. In the case of equations, the preliminary schema takes longer to build.

## TWO KINDS OF AUTHORITY

Whenever, and to the extent that, ideas prerequisite for understanding have not been made available to the learner, then whatever is communicated can only be in the form of assertions, and these will not provide nourishment for a growing intelligence. (The food metaphor is a close one. Genuine nourishment becomes part of the bodily self of the person who eats it; indigestible material is internalized, but not assimilated, and efforts to retain it indefinitely are contrary to our natural functions.) The acceptance of an assertion depends on the acceptance of the teacher's authority, and acting on it partakes more of the nature of obedience than of comprehension. In contrast, the assimilation of meaningful material depends on its acceptability to the intelligence of the student. Acting on it results from, and consolidates, enlargement of the learner's schemas.

So far the word 'authority' has been used in what is probably its commonest connotation, that of a person to whom respect and obedience are due, as a result of status or function. Authority can also, however, result from superior knowledge, and this is, or should be, the kind of authority pertaining to a teacher as such. In schools (where we make our first, and some of us our last, attempts to learn mathematics), however, there is confusion and conflict between these two kinds of authority.

The former is closely related to the establishment and maintenance of discipline—of orderly behaviour and obedience to the teacher's instructions. This is the same kind of discipline, though of a milder kind (usually), than that imposed in the armed forces. But we talk also, though less commonly, of the disciplines of mathematics, chemistry, philosophy, etc. When a great scholar attracts disciples, they come as learners, and when they obey, it is willingly, because they want to learn.

School teachers have to exercise both kinds of authority, and promote both kinds of discipline. If they fail to control their young pupils[1] who do not attend school of their own free will, they have little chance of teaching them. Yet basically these two roles are not only different but in conflict. In other circumstances they are usually separated. At a meeting of a learned society, the former role is exercised by the chair, who calls the meeting to order, indicates whose turn it is to speak, and in general controls the conduct of the meeting; the other speakers—invited guests or participators in the audience—inform and discuss. It

---

[1]Reminder: as I use the word (see footnote on page 4), a pupil is someone who attends school (etc.) under compulsion. This may have a major influence in the learning situation. See Skemp (1979a), all of Chapter 15.

is improper for anyone to act contrary to the authority of the chair, but entirely proper for anyone to question and discuss the remarks of any of the other speakers, however eminent.

The combining of both these functions in one person may be necessary, but it is certainly unfortunate. In matters of orderly behaviour, performance of assigned tasks, choice of subject-matter, it is my view—which some will consider old-fashioned—that children should accept the controlling role of their teachers; whereas the learning with understanding of the subject-matter thrives on questions and discussions among learners and between learners and teacher. Usually a reasonably satisfactory *modus vivendi* is reached, in which pupils learn how far teachers, in their first role, allow and even encourage them to express disagreement with them in their second role. Even so, artful pupils may use the second as a disguised form of opposition to the first, while teachers may subjectively experience a genuine request for explanation as a questioning of their (controlling) authority, and react inappropriately.

This role conflict matters particularly in mathematics for the reasons given earlier, that for this of all subjects, its learning and teaching need most to be based on reason and agreement. The situation will be aggravated whenever teachers are unable to give good reasons, because (perhaps through no fault of their own) they do not know them, and whenever (for lack of an adequate conceptual analysis) they have not developed the learners' schemas in such a way that the material is experienced *by them* as reasonable. In these conditions learning based on understanding breaks down and is replaced (if at all) by learning based on respect and obedience.[2]

## BENEFITS OF DISCUSSION

So far we have centred our attention on the teacher-learner relationship. But discussion with fellow-students can also be an important contribution to learning. The mere act of communicating our ideas seems to help clarify them, for, in so doing, we have to attach them to words (or other symbols), which makes them more conscious. 'A problem clearly stated is half solved', and we have all found on occasion that in the process of formulating some problem, personal or academic, to a willing listener, we ourselves arrive at a solution. I met a teacher who uses an interesting technique when, in discussion, learners make misstatements. A common response is to ask another learner to explain to them where they are wrong. This teacher, however, asks learners to explain to the rest of the class the

---

[2]I wish to make clear my personal view (with which some may disagree) that, in their appropriate spheres, respect and obedience are necessary and desirable, since, without these, the conditions necessary for learning of the other sort could cease to exist. What I am trying to clarify is the distinction between two kinds of learning situation.

reasons for their statement. The usual result is either that they discover their own error after a few sentences, or that the rest of the class learns something new.

But there is more to a discussion than just thinking aloud. Another factor is the interrelating of our ideas with those of others—the expansion of our own schemas to enable us to assimilate their ideas, and the explanation of our ideas to them to enable them to assimilate our ideas to their schemas. Both are demanding, in different ways. The former requires flexibility and open-mindedness; the latter requires the ability to see just where the differences between one's own schema and that of the learner ('to see things from the other's point of view') lie, in order to know what explanation is necessary to bridge the gap. But if we can meet these demands, our own schemas will become enlarged thereby. More important still, they become more flexible: that is to say we, as total personalities, acquire habits and attitudes which favour further growth of our schemas.

Discussion also stimulates new ideas. One factor is simply the pooling of ideas, so that the ideas of each become available to all. Imagine a jigsaw puzzle in which the pieces are distributed among several persons, none able to see those of the other. Each might be able to complete part of the puzzle, or the pieces of each might be quite disconnected. But spread the pieces out on a table where everyone can see all the pieces, and they can all work together at fitting them together to form a meaningful whole.

The cross-fertilization of ideas is another benefit which comes from discussion. Listening to someone else (or reading what they have written) may spark off new ideas in us which were not communicated to us by the other, but which we would not have had without their communication. These may then, in turn, spark off new ideas in them, the result being a creative interaction which, at its best, can be exhilarating to all concerned.

Probably the best numbers for a creative discussion of this kind are small—two only, or at most three. Sometimes a new and fascinating idea is evoked, but before one can grasp hold of it the other person says something else, unwittingly distracting one's attention, and the fleeting glimpse is lost. A friend has suggested to me that there should be signals in discussion whereby either party may ask for silence if it is needed. Pencil and paper would also help the future retrieval of the idea and allow talk to resume. In this way a situation would be established unifying public discussion, private thought and written notes. This suggestion requires, and describes, good personal relationships between those taking part, and so points the way to another aspect of discussion.

## Attitudes Within Groups

These benefits of discussion are also very dependent on friendly and fairly informal personal relationships between the members of the group. This does not mean quite informal. There must be certain agreed forms of behaviour, such as a willingness to take turns to speak, to listen, to consider the viewpoint of others.

These are important parts of civilized discussion and are not too easily achieved. We do not much notice these forms of behaviour, because they facilitate the main task and do not intrude themselves on it.

If we do not like the fellow-members of our group, we are unlikely to be interested in sharing ideas, relating our own to theirs, or looking at things from their points of view. Rather the opposite—both we and the others will, according to temperament and circumstances, either try to make the group conform to our own ways of thinking or insulate ourselves from the pressures of the rest of the group.

This does not mean that the members have to agree in all their ideas or viewpoints; it means that they have to disagree in the right kind of way. That is, they have to agree that they will conduct their discussions on a rational basis and will neither make, nor react to, attacks on their statements or arguments as if these were attacks on themselves. And they have to agree on the final goal of any discussion—a step forward, by all, in the understanding of the subject.

## The Teacher as a Group Leader

An attitude such as that described above is a very mature one, which by no means all who take part in discussions achieve, whether children, adolescents or adults. And we also know, all too well, that people in groups can be much *less* creative, more destructive, sometimes even less human, than their members individually. Indeed, this seems to happen more easily than the creative interaction we have been discussing.

Which factors are at work is not yet fully known. Research on Freudian lines suggests that some of these are unconscious. Two of them, however, are clear— size and leadership. A large group degenerates into a mob more easily than a small one, and the part which each individual can take in discussion diminishes rapidly with size. My own experience is that quite small groups, numbering from two to five or six, are best, and although thirty to forty is a more usual number for a school class, there is currently a trend, particularly in primary schools, towards either individual work or work in small groups.

Where traditional class teaching is used with a fairly large class of pupils, i.e., those who are not there by their own choice, there is a situational pressure on teachers to take up an authoritarian attitude. If they do not establish and maintain order, they cannot fulfil their function as communicators of knowledge. Nevertheless, these two roles are basically in conflict, as indicated earlier, and the larger the group, the greater the conflict.

Ideally, a good school-teacher has to be both sergeant-major and conductor of an orchestra, able to alternate between these roles as required. To combine this with a knowledge of the subject is asking a great deal. I once watched a lesson in which the teacher achieved all three. Her control of the class was so well established and effortless that this role was not noticeable at all. In the course of the lesson a girl gave a wrong answer. The teacher wrote it on the blackboard; then by skilful questioning she led the class as a whole not only to find the right answer but to learn more from the wrong answer than they would have if the first answer had been correct. Moreover, the girl who gave it was not made to feel ashamed of, or embarrassed by, her mistake. It was also interesting to sense the intra-group feeling at the stage when about half the class had seen the point while half had not. Those who did understand showed, on their faces, the pleasure which accompanies a new insight; but they were also genuinely concerned to try to 'pull the rest of the class over the stile'. When everyone understood, there was a general relaxation of tension and feeling of satisfaction. The handling of her class by this teacher so impressed me that, at a subsequent meeting of teachers, I asked her to tell us how she did it. After a few minutes it was clear that she did not know, consciously. Her skilled group leadership functioned at the intuitive level and not yet at the reflective level.

Those who really understand mathematics are not common; those who can communicate it, less so; those who are also excellent group leaders, fewer still; while those who can also communicate this last ability are rare indeed.

## ANXIETY AND HIGHER MENTAL ACTIVITY

Another reason why the right kind of interpersonal relationship is so important in understanding mathematics is that anxiety itself may increase, subjectively, the difficulty of understanding. Given an exposition which, though not excellent, is more or less adequate, some learners will be able to understand it, some not. If those who do not understand feel over-anxious at their failure, they will no doubt make greater efforts to comprehend. But this over-anxiety can be self-defeating, in that it can actually diminish the effectiveness of their efforts. The more anxious learners become, the harder they try, but the worse they are able to understand; and so, the more anxious they become. Thus a vicious circle may be set in operation. Indeed, two: the short-term one just described and also a long-term one. Given several experiences of this kind, the situation itself, the mathematics lesson or lecture, becomes a learnt stimulus for anxiety; so the learner begins each lesson already partially defeated. That this is not an exaggerated picture will be vouched for by many from their personal experience.

Here are some arguments in support of the belief that anxiety reduces—or may in certain conditions reduce—efficiency of mathematical thinking.

A principle known as the Yerkes–Dodson law has now, on the basis of experimental evidence, been fairly generally accepted by psychologists. This law states that the optimal degree of motivation for a given task decreases with the complexity of the task. In other words, for a simple task, the stronger the motivation the better the performance. But for a more complex task this is only so up to a point. Starting from zero motivation, which presumably produces zero performance, increasing the motivation improves the performance. But beyond a certain degree of motivation, further increase produces no further improvement of performance, but a deterioration. And the more complex the task, the lower the degree of motivation which gives the best performance.

Motivation is a fairly tricky thing to assess accurately, though performance is usually straightforward. This is because motivation is internal to the person concerned and not directly observable, while performance is observable and can be objectively assessed. To assess motivation experimentally, we have to set up conditions which we assume will have certain motivational effects on the subjects. For example, in one experiment rats were required to solve discrimination problems under water. They were confronted with two different doors, one of them locked, the other open and leading to air. The level of motivation was here varied by keeping them submerged for 0, 2, 4 and 8 seconds before they were allowed to start. Three different levels of difficulty of problems were used, and the results were in accordance with the Yerkes-Dodson law.

Understandably, there is less evidence of this kind available concerning human subjects. But let readers imagine themselves in a field when they discover that a bull is advancing menacingly upon them. The fiercer and closer the bull, the better their performance will be at running (a task of low complexity), jumping a ditch or climbing a gate. But suppose that the bull breaks through the hedge and readers seek safety in their car. Then, in the slightly more complex task of finding the right key and unlocking the car, they might well fumble. If the key were not in its usual pocket, they might take longer to remember that they had hidden it where others of the party could also find it if they returned first. Or suppose, by a stretch of the imagination, that they had to solve an easy problem to escape (as did the experimental rats), readers might well find that they took longer to do this than they would have under more relaxed conditions.

That the higher mental activities are the first to be adversely affected by situational anxiety has long been recognized by the army. Actions which have to be done under battle stress are taught as strongly formed habits, to be performed automatically, while those who have to plan the strategy of the battle and direct its tactics are kept out of the firing line. Many teachers, recognizing that examinations are a stress situation, similarly drill candidates in well-practised routines.

My own experiments in this field have been based on the hypothesis that it is the reflective activity of intelligence (see Chapter 4) which is most easily inhibited by anxiety. One task used to test this hypothesis was a simple sorting task. Cards were prepared having one, two, three or four figures of the same kind on each. These figures could be squares, circles, crosses or triangles, and they were either red, green, yellow or blue, all figures on a card being alike and of the same colour. Four category cards were laid out: one red triangle, two green squares, three yellow crosses, four blue circles. The subject was given the remaining sixty cards and asked to sort them into piles in front of the category cards according to a single criterion. For example, a card having four green crosses would, if the subject had been told to sort by colour, be placed in pile two from the left. If sorting were by shape, it would be put in pile three; if by number of figures, pile four.

When the same criterion was used throughout, the subjects performed the task rapidly and efficiently. Moreover, their speed increased with practice. Subjects were then asked to sort the first card by colour, the second by shape, the third by size, the fourth by colour, and so on. This was no longer a routine task, but one involving reflective activity, albeit of a simple kind. (The reader may find it helpful to refer back to the diagram on page 38.) The subjects had to be aware of the category in use, this category being something internal to their own minds, not something external, and they had to switch this category to the next in series after each card had been sorted. The first is a receptor, the second an effector, activity of the reflective system.

The subjects were asked, as for the first task, to sort as fast and as accurately as they could. But under these conditions, far from improving with practice, they got steadily worse. Sometimes they broke down altogether—that is to say, they suffered a sort of 'mental blockage' during which time they could make no progress at all with the task. One subject, a university student of high intelligence, reported 'waves of panic, which I had to fight back.' The subjects were aware that they were being timed and that their errors would be noted, but were not otherwise subjected to external stress. It was quite striking how changing the task from a routine one (after a single reflective act at the beginning, to 'set up' the chosen sorting category) to one involving continual reflection was sufficient to produce conditions in the subject leading to moments of apparent mental paralysis.

It seems possible that the progressive effect might be due to the vicious circle described earlier. The worse the subjects performed, the harder they tried, and so the worse they performed, with consequent mounting anxiety. If this hypothesis were correct, then the interpolation of a simple routine task would interrupt the cumulative effect, and performance at the reflective activity would improve. This hypothesis was tested in a group experiment[3] with fifteen-year-old grammar-school boys, and it was found that the progressive decline in performance was in fact removed.

Most of us can recall occasions when we have experienced a similar kind of momentary mental blockage. After important interviews, perhaps we feel that we could have given a better account of ourselves. Remembering that an interpolated routine task helps to reduce anxiety, I often, when interviewing candidates for university admission, begin with very straightforward questions and then

---

[3]Details of the first experiment have been given in full, since it is an easy one for readers to repeat if they wish. The tasks for the group experiment were different, and for reasons of space are not given here.

interpolate more of these at intervals throughout the interview. Similarly a good teacher can, by initially asking questions that the learner can answer, reduce anxiety and build up confidence, and thereby improve the performance; a bad teacher can reduce an averagely intelligent pupil to tongue-tied incompetence.

Here we are back to the interpersonal situation, and when considering the learning of mathematics it is difficult to keep away from it for long. Even adult students, learning independently from a text, cannot escape the historical effects of early teachers on their own self-confidence, or lack of self-confidence, in the mathematical-learning situation. When teaching elementary statistics to psychology students, I found that with many of these my first task was a remedial one, to convince them that they were indeed able to comprehend mathematics. I hope that readers who have unhappy memories of past attempts at learning mathematics will be willing to accept, as a working hypothesis, that the causes were other than their own lack of intelligence.

## Initial Causes of Anxiety

In the last section it was suggested that anxiety, once present, could bring about a vicious circle of cause and effect in the mathematical-learning situation. On the principle that prevention is better than cure, we should now look for the causes which may introduce anxiety in the first instance.

One of these has already been suggested—an authoritarian teacher. This includes, of course, the strict disciplinarians of the old school. But in a milder sense, we must also remember that whenever the schemas necessary for comprehension are not present and currently available in the mind of the learner, whatever learning takes place can only be based on an acceptance—a willing acceptance, perhaps, if the teacher is well-liked—of the teacher's authority. Learning of this kind is rote-learning not schematic-learning. Initially it may not be accompanied by anxiety, perhaps quite the opposite. Well-memorized multiplication tables, resulting in a column of neat red ticks, are rewarding to teacher and pupil alike. The problem here is that a bright and willing child can memorize so many of the processes of elementary mathematics so well that it is difficult to distinguish it from learning based on comprehension. Sooner or later, however, this must come to grief, for two reasons. The first is that as mathematics becomes more advanced and more complex, the number of different routines to be memorized imposes an impossible burden on the memory. Second, a routine only works for a limited range of problems and cannot be adapted by the learner to other problems, apparently different but based on the same mathematical ideas. Schematic learning is both more adaptable and reduces the burden on the memory.

Learners of the kind described therefore inevitably reach a stage at which their apparent success deserts them. Try as they may, they can no longer 'get all their sums right.' The efforts they make are, of course, along the wrong lines—of trying to remember more and more rules and methods. Really they need to go back to the beginning and start again on new lines. Were this possible, the well-learnt routines would stand them in good stead (see page 61). But neither they nor their teacher know what is the matter, and even if they did there would probably not be time.

Here indeed is an anxiety-provoking situation, and there are now two vicious circles likely to be set in operation. The first is that described in the last section; the second is that the increasing efforts the learner makes will inevitably use the only approach which he knows, memorizing. This produces a short-term effect, but no long-term retention. So further progress comes to a standstill, with anxiety and loss of self-esteem.

## Adaptations to Anxiety

Two important qualifications must now be made to the foregoing. The first is that the Yerkes-Dodson law refers to motivation in general, and we have so far been concentrating on motivation by anxiety. This is by no means the only, or the best, motivation. The second is that the optimal level of motivation for a given task will depend on individuals as well as the task. This is to some extent implicit in the earlier statement, that the optimal level decreases with the complexity of the task, for what is a complex task for one group of people may be a relatively straightforward one for another. The greater competence of the latter will help them in two ways: they will feel less anxious because they know they can cope; and they can use their anxiety constructively at work on the problem. A certain amount of anxiety can be a useful stimulus, and part of the background of education is to learn to use it as such. This we may call 'adaptation to anxiety.' Part of this adaptation results from having techniques for resolving anxiety-provoking situations—solving the problem or passing the examination in our present context. But another part is a personality variable. As such, it falls outside the scope of this book, but it is well worth noting that many of those who have contributed to knowledge have been not without their personal problems. We may speculate—but at present it is only a speculation—that in some way these people have found in their work an escape from anxiety, perhaps because it is an impersonal situation, perhaps also because it contains problems which they can solve.

## MOTIVATIONS FOR LEARNING

So far our efforts have been directed towards trying to understand some of the factors which affect the learning and understanding of mathematics, on the assumption that we, or the learners concerned, want to do so. But now, let us in all seriousness ask the question: why should anyone want to learn mathematics?

It is indeed arguable that this question should have come right at the start of the inquiry, since without some kind of motivation there would be no reason to expect anyone to make the necessary effort. However, if you have bought this book, you probably have some kind of motivation. So let us now look at what this might be—or these, since several motivations may combine in a single activity.

'Motivated' is a description we apply to behaviour which is directed towards satisfaction of some need. If we say that a certain piece of behaviour seems motiveless to us, we mean that we do not know, and cannot even guess, what need is satisfied by means of it. So questions about motives are usually, in disguise, questions about needs.

Some needs, such as food, warmth, sleep, are innate. Others, such as tobacco, soap, a television set, are learnt. Mathematics seems fairly obviously to be a learnt need; so how (if at all) do people learn to need mathematics?

One way in which new needs can be acquired is by learning that they lead to the satisfaction of already existing needs. In our present culture we soon learn that if we have money we can use it in many different ways to satisfy a wide variety of needs. Mathematics is also a valuable and general-purpose technique for satisfying other needs. It is widely known to be an essential tool for science, technology and commerce, and for entry to many professions. These are goals which motivate many adults to mathematics, but they are too remote to be applicable to the early years of school, when we first begin mathematics. In the classroom, shorter-term motivations are more likely to be effective: two of the most directly applicable here are the desire to please the teacher and the fear of displeasing her or him. Reward and punishment are widely used as methods for training both children and other young animals, and are older than schools themselves.

Both of these kinds of motivation are extrinsic to mathematics itself, however. Teachers can be pleased, or their displeasure avoided, by emitting the desired behaviour (verbal or written) with little or no understanding, so there is no guarantee that understanding has been achieved. Indeed, since understanding may take longer than parrot-learning, extrinsic motivation of either kind may favour the latter because it brings prompter results—quicker approval or quicker relief from anxiety, as the case may be. Of the two, motivation by anxiety is probably the more conducive to rote-learning, because, as we have already seen, it has an inhibitory effect on the reflective activity of intelligence.

## Intrinsic Motivations

But there are some people for whom mathematics is a pleasurable and worthwhile activity in itself, regardless of any other goals which it may also serve. These are the people whom I regard as true mathematicians, and if this view is accepted, then some seven-, ten- and twelve-year-olds merit the description as much as many sixth-form and adult students. Why people should enjoy learning and practising mathematics for its own sake is, however, far from obvious if we keep to our original hypothesis that any motivated behaviour satisfies some need.

Let us approach the problem indirectly, by way of other examples. Look at a child, out for a walk with parents, balancing along a low wall in preference to walking along the pavement. Or look at a dinghy sailor, sitting precariously out over the water in preference to the greater certainty and convenience of an outboard motor. Or look at a mountaineer, laboriously and hazardously climbing a mountain, which he could ascend quickly and safely by a funicular railway. Wall-walking, sailing, mountaineering are not basic needs, but neither do they appear to be used as means to other goals, since in each of these examples there is a simpler and more direct means of attaining the end.

The apparent contradiction can be resolved if we hypothesize another very basic, very general need—a need to grow. The word 'grow' is used here to include not only physical growth but growth in skill, power, knowledge and any other physical, sensori-motor or mental organization which actually or potentially favours survival. Young children do not need to balance on walls, climb trees, jump off climbing frames, do forward rolls. But all of these serve, very directly, their growth needs: they develop the lungs, muscles and bodily control.

Mental growth is even more important for survival than physical growth, and activities which contribute to mental growth should therefore be enjoyed by children at least as much as physical activities. Mental growth, moreover, can continue long after physical growth has ceased, so the pleasures which come from the various ways of exercising one's intelligence should continue from childhood to old age. If it is agreed that genuine mathematics is simply a specialized form of intelligent activity, then we need no longer wonder why it can be enjoyable for its own sake.

The enjoyment we experience from activities, physical or mental, which serve our growth needs are experienced as intrinsic in the activity itself. Children don't like climbing because they know that it makes them strong and agile. Rather, they grow strong and agile because they like climbing. What is more, letting children climb trees is a better way to help them become strong and agile than making them do exercises. The rewards of doing something one enjoys are immediate and conducive to prolonging the activity itself; the more distant the goal, the greater the imaginative span required to relate present activities to it, the slower the apparent progress—in relation to the whole distance to be traversed—and, in general, the weaker the motivation.

For an adult, an excellent learning situation is one in which short-term and long-term motivations are fused, the short-term one being an enjoyment of the learning and doing of mathematics—an intrinsic motivation—and the long-term one being some personal, practical or academic goal to be achieved with the help of a knowledge of mathematics. But of the two, intrinsic motivation is probably the more important. Some things we learn because we know that they are useful.

But the major strides which have been made, in mathematics as in the sciences, have resulted from the quest for knowledge for its own sake. Faraday is said to have replied to a woman who saw his demonstration of the deflection of a compass needle by a coil of wire through which an electric current was passing and asked what use that was: 'Madam, of what use is a new-born baby?' One characteristic of a baby is that it will grow, and another is that we cannot predict into what kind of adult. Even Faraday (1791–1867) could hardly have guessed at the long-term results of his discovery, whereby the connection between magnetism and electricity was established.

Similarly, a tendency towards growth is an intrinsic quality of the kind of mental organization which we call a schema. That we experience pleasure from any activities which are favourable to their growth is the most powerful incentive to learning, mathematics or any other subject. That the knowledge will afterwards be useful, or in what way, cannot be predicted at the time of learning, any more than when I buy a screwdriver I know exactly what jobs I am going to do with it. When they were studying calculus and algebraic geometry in college, the mathematicians of the American space research programme did not know that they would be using their knowledge to plot orbits for a lunar module.

How effective an intrinsic motivation for learning mathematics can be is something which many teachers do not yet appreciate. On a number of occasions, teachers finding that children actually enjoy mathematics when it is intelligently taught and learnt have reported this to me with a mixture of surprise and pleasure; but also of doubt, as if something must be wrong with an approach to mathematics which children enjoyed. But until this intrinsic motivation is better comprehended and put to work, mathematics will remain for many a subject to be endured, not enjoyed, and to be dropped as soon as the necessary exam results have been achieved.

# PART B

# 8

# A New Model of Intelligence

---

## TEACHERS AND LEARNERS

A human child is at the most learning age of the most learning species that has yet evolved on this planet.

We learn in a variety of ways. Some of these, such as operant conditioning and habit learning, we share with other animals. But one of these ways is unique to man—to homo *sapiens:* not just the naked ape, as a zoologist (Morris, 1967) has called us, but the understanding ape—the ape with the potential for knowledge, even wisdom. The extent to which this potential is realised will depend, for each individual, on how he develops the intelligence with which he is born, just by virtue of being human.

This, in turn, will largely depend on his teachers. These begin with his informal teachers, such as his parents, grandparents, older brothers, and sisters. Soon, and importantly, these come to include professional teachers.

Some of the best teaching is indirect (e.g., by providing interesting activities for children to explore together). So by teaching, I mean any kind of action that is intended to influence the learning process. This process is inaccessible to direct observation by an outside person in the same way as our internal bodily processes are to a medical practitioner. In both these cases, a person who intervenes without an adequate mental image of what is going on inside is as likely to do harm as good.

In this chapter, I offer an outline for a mental model of intelligence. A full presentation occupies a 300 page book (Skemp, 1979a), so a single chapter in the present book can be no more than an outline. However, 'a journey of a thousand miles begins with a single step', and in this chapter I offer some suggestions for the first few steps.

## HABIT LEARNING AND INTELLIGENT LEARNING

I begin by developing further the distinction between habit learning and intelligent learning.

Habit learning has been extensively studied by the behaviourist school of psychologists, mainly with animals. A well-known example is provided by the Skinner box. A hungry rat is put in a cage, in which there is a bar sticking out horizontally from one side. In the course of its movements around the cage, the rat happens to press the bar, and a morsel of food is released into the cage. Eating the food reduces hunger, and each time this happens, the association between the stimulus situation (being in the cage, hungry) and bar-pressing is reinforced. Gradually this builds up into a habit.

This kind of learning is also found in humans. Here is an example. A little boy's parents noticed that he had acquired a new and puzzling behaviour when given a biscuit. This was, while eating it, to hold the biscuit above his head between bites. When asked why, he didn't know. On reflection, his father remembered that while on holiday they had been friendly with a family who had a little dog called Penny. Penny also liked biscuits! So we now had a probable explanation of this habit, which had persisted even when it was no longer required.

In habit learning, certain actions are reinforced as a result of their outcomes, so learning *follows* action. And what is learned *is* action; the cognitive element is small. Rote learning, such as we do when we memorize a telephone number, is verbal habit learning.

Once learned, habits are very persistent. They have low adaptability: the little boy went on with his habit long after it was no longer useful, and if our telephone number is changed, we cannot erase the old number from our minds as we do from our desk pad. The old number persists, and gets in the way of the new one.

In contrast, the main feature of intelligent learning is adaptability. Our actions are goal-directed, not stimulus-determined. We use flexible plans of action, adaptable to each new situation. These can be constructed in advance of action, and modified in the light of action. They enable us to achieve a wide variety of goals, in a wide variety of situations. What is more, we can devise several plans, and choose the best, before putting this plan into action. Intelligent learning often *precedes* action. And action is used not only for achieving goals, but for testing hypotheses.

At the level of habit learning, one can find some justification for the behaviourist view that our behaviour is shaped by the environment. In sharp contrast, my model for intelligent learning asserts that we can also shape our own behaviour, to achieve the same goals in different environments. If I am thirsty and want a drink of water, at home I go to the kitchen and turn on the tap. As a child, I went to the yard and worked a pump handle. In a cafe, we ask a waitress. In camp, we find a clear stream. On a winter day when our pipes froze, our breakfast coffee was made with melted snow. Different environments, different plans of action, all directed toward the same goal: relief of thirst. Each plan is based on knowledge of the environment; and building up this knowledge is a major function of intelligence. Action is not a response to an external stimulus, but directed toward whatever goal an individual has in mind. The cognitive element is great, and as a result, so is the variety of available plans. *Knowledge gives adaptability.* Science and technology are more sophisticated examples of this combination, of well-devised plans of action derived from knowledge. Science is concerned with building up knowledge, technology with applying it. Math is important for both.

Both kinds of learning are available to us as humans, and both have their uses. In learning to spell, rote learning is necessary for words spelled and pronounced irregularly, such as bough (on a tree), bow (respectfully), bow (and arrow), trough, enough; although we can use intelligent learning for words whose spelling is regular, such as can, ban, fan, man. The mix is different for different subjects. The right mix for mathematics is about 5% rote learning, 95% intelligent learning: but all too often it is taught in a way which makes the latter impossible.

## WHEN IS A STIMULUS NOT A STIMULUS?

At the end of my road there is a red mail box. When I have a letter to post, in behaviourist terms this box is a stimulus to which I respond by putting the letter through the slit. From the main road my side road is inconspicuous, so when I am returning home in my car from shopping downtown, this red box acts as a landmark. Now it is a 'stimulus' for a different response: to signal a right turn, slow down my car, look carefully both ahead and in my mirror, and either turn or stop until it is safe to turn right.[1] When returning from the opposite direction it is now a stimulus to turn left, which in this case requires less caution. And when I am going from Coventry to Kenilworth, this same stimulus evokes no response—in effect it ceases to be a stimulus.

The burden of the foregoing will be apparent. I do not find the notion that our behaviour is stimulus-determined a useful one in the context of intelligence. It belongs, if anywhere at all, to the levels of automaticity in our behavior.

---

[1]Extra care is needed in this direction because I am turning across oncoming traffic. In England we drive on the left!

## GOAL-DIRECTED ACTION

It is a matter of simple everyday observations such as the foregoing that much human activity is goal-directed.

This implies that if we want adequately to understand what people are doing, we need to go beyond the outward and easily observable aspect of their actions, and ask ourselves what is their goal. To say that Richard Skemp is riding his bicycle may be perfectly true, but it is only part of what matters. I may be riding to my office; or I may be enjoying physical recreation with my son; or I may be testing the adjustment of my three-speed gear.

Moreover, taking the first goal state, this may underlie actions that look quite different. For example, instead of bicycling I might be standing quite still by the side of the road, waiting for a bus; or I might be walking (which in this case would also be along a different path); or I might be driving my car (in which case I would start off in a totally different direction). To limit a description of what was happening to the observable behaviours, superficially very different, would be to miss what they had in common, namely the goal state. I am not claiming that all our behaviour is goal-directed; only that some, perhaps much of it, is; and that where this is the case, to ignore it is to ignore one of its most important features.

## SURVIVAL, ADAPTABILITY, AND EVOLUTION

If this assumption is true, one question leads to another. If we can identify the goal underlying certain observable activity, we want next to know what is the significance of that goal rather than some other. And the more examples we look at, the more it becomes apparent that many of our goal states have one thing in common: that they favour survival.

In the bicycling example, three possible goal states were mentioned: being in my office, physical recreation with my son, and checking my three-speed gear. Among the properties of the first goal state are being physically in a good position for doing my job—having near me colleagues, students, a telephone, books, reprographic facilities—and thereby (amongst other things) earning a living. My salary enables me to pay for food, clothing, shelter; and without these, I would not long survive. Physical recreation with my son has health-giving properties for both of us, whereby we are both likely to live longer; and it is through our children that our families, our nations and our species continue to survive when we do not. An efficient three-speed gear enables me to reach my destinations more easily, and so makes makes available more time and energy for other purposes—again, indirectly, with benefit to survival. Medium-term, long-term, and short-term, all three choices of goal state, and success in reaching them, have properties that contribute to survival.

And for a large proportion of the goal states we choose, we can find, once we get in the way of looking for them, properties that relate favourably to survival: be this state physical location, some other physical state such as temperature, or a mental state. Often these are interconnected. A person who is lost on the moors in winter (not in the mental state of knowing where he is) cannot reach a physical location where he will find shelter and warmth: and he may die as a result.

This frequent relation between our choice of goal state and survival should come as no surprise. Every species now in existence is here because, and only because, it has evolved characteristics that enable it to survive. These characteristics are both physical and behavioural. Most of the former, and some of the latter, are genetically programmed: they form part of the survival equipment of the species. But the evolutionary process of adaptation of these innate characteristics, although often very effective, is slow. If the environment changes faster than a species can adapt, that species will become extinct. Under these conditions, one of the most valuable characteristics that a species can evolve is not a visible one, either physical or behavioural: but the invisible asset of adaptability. We, as members of the species *homo sapiens,* are the most adaptable species of any; and what gives us this adaptability is an ability to learn in a particular way, which I call *intelligent learning.* The ability itself I call *intelligence.* To the relations between intelligence, adaptability, and the achievement of our goals, we now turn our attention.

## DIRECTOR SYSTEMS

The starting point of this model was the assumption that many, perhaps most, of our actions are systematically directed toward bringing about goal states. Now, the situations in which we find ourselves doing this are not always the same. Nevertheless, we are able to achieve our goals by varying our actions appropriately, and this ability is in itself very pro-survival. A system for doing this I call a *director system,* and because the idea is borrowed from cybernetics, the ways in which a director system works will be familiar to many readers. But I need to recapitulate them briefly for the sake of what comes after.

Its essence is a comparison between the present state of some operand in the environment and its goal state, combined with a plan of action directed always so as to reduce this difference until the present state coincides with the goal state. This means that both the present state and the goal state have to be represented within the director system in some way: otherwise, how can they be compared? So, in slightly greater detail:

We need a *sensor,* which takes in information about the present state of the operand, and represents it internally. In Fig. 8.1 it is represented by an eye. We need an internal representation of the *goal state.*

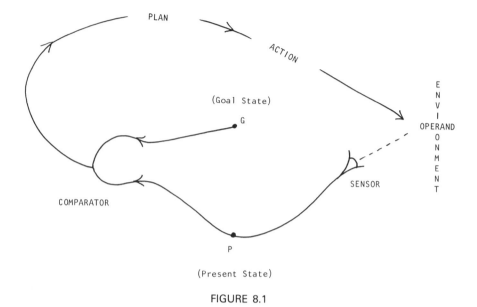

FIGURE 8.1

We need a *comparator,* to compare these.

And we need a *plan of action:* what we actually do to change the state of the operand from its present state to the goal state.

The distinction between the possible sources of these plans of action, and their different resulting natures, is a key feature of the present model of intelligence.

## GOAL-DIRECTED LEARNING

In the lower animals, many of their director systems are innate, complete with instinctual plans of action. These form part of the survival equipment of the species. But there is an upper limit to what can be transmitted genetically; and there are other disadvantages, such as slowness to adapt to environmental changes. So it is not surprising that some species have evolved the ability to develop new director systems during the lifetime of an individual, and to improve the ones they have. This is how I now conceptualise learning: as a change in an organism's director system toward a state of better functioning. Other animals can learn too, of course. But we have so far outdistanced other animals in our ability to learn that I regard our own species as an evolutionary breakthrough.

It is not so much that we can do better than other animals at the same kind of learning. At maze learning, for example, rats are better than humans. But we have available a more advanced kind of learning, which is qualitatively different

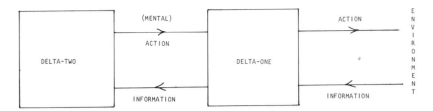

FIGURE 8.2

from the kind exemplified by maze learning. It is the ability to learn in this more advanced kind of way that I now call *intelligence*.

Intelligence is a kind of learning that results in the ability to achieve goal states in a wide variety of conditions, and by a wide variety of plans. And if we have been able to agree that action (much of it) is not stimulus-determined but goal-directed, it may be acceptable to take a further step: which is to conceive that there may be a kind of learning that is also goal-directed.

In the present model of intelligence, not only is action conceived as goal-directed, but learning itself: In this case, the building and testing of integrated structures of knowledge. For this we need two director systems, which I call *delta-one* and *delta-two*.

Delta-one is a director system acting on operands in the physical environment. Some of its structure has been shown in Fig. 8.1. This detail has been omitted in Fig. 8.2 for the time being.

Delta-two is a second order director system that has delta-one as its operand. Its function is to take delta-one to states in which delta-one can do its job better. These new states are goals of learning.

So now let us consider what these states might be.

## GOALS OF INTELLIGENT LEARNING

In order to perform its function, of getting something from its present state to a particular goal state, delta-one has to have a plan of action. Here is one that I was given some time ago. I had followed, in my own car, that of the local maths advisor to a school where I would be working. Afterward, he gave me the following directions for getting to highway A45, after which I could find my own way home.

LEFT
LEFT
LEFT
BEAR RIGHT
STRAIGHT ON
LEFT

You can guess what happened. After five minutes, I was hopelessly lost; and it was not until I had wasted much time that I eventually found highway A45 and was heading for home.

A plan of this kind is closely tied to action. And if just one of the actions is wrong, it is liable to throw out all which follow. In this case, if one gets off the correct route one doesn't know what to do to get back on. The cognitive element is low. Habit learning is of this kind: if I had memorised the foregoing plan correctly, or even written it down, the above disadvantage would have remained, together with the additional ones to be described later.

For my next visit, I made sure that I had a street map (see Fig. 8.3) with me. On this, I could identify the particular route I had been told last time.

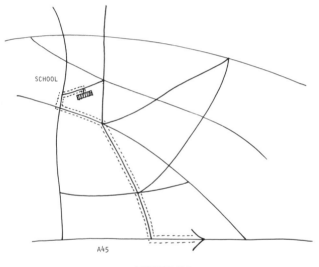

FIGURE 8.3

I could see from this where I had gone wrong—I had not counted the turn out of the school drive as my first left turn. But now, even if I took another wrong turn, all I had to do was to find out where I was by looking at a road name, after which I could both find where I had gone wrong, and work out a plan for getting back on my route.

This simple example highlights some of the chief differences between a fixed plan, which was all I had on the first occasion, and a knowledge structure, such as I had on my second visit. From a knowledge structure one can if necessary find out what one has done wrong. One can correct one's mistakes. And there is a less obvious, but even more important, distinction. This is in the economy and adaptability of the second kind of learning. This particular knowledge structure took the form of a map. Given any present state (in this case, location), and any goal state, one can devise a plan of action (in this case a route) to take the operand (in this case myself) from the first to the second: provided only that both can be represented on the same map. This is a much more effective and economical way of learning because it gives the potential for devising a very large number of plans, which to remember separately would impose an almost impossible burden on the memory. Even in the small road network shown in Fig. 8.3, if we regard each stretch of road between two junctions as a single location, and each intersection as another location, the number of locations is 56. The number of possible combinations of present state and goal state, is 3,080. To learn this number of separate and disconnected plans would impose an impossible burden on the memory. But the knowledge structure represented contains the potential for each and every one of these.

Using as a transitional metaphor Tolman's useful idea of a cognitive map, this example generalises very nicely into the concept of a knowledge structure, for which another name is *schema*. One of the main goals of delta-two is the construction of these schemas.

It is learning of this kind with which intelligence is particularly concerned; and though its function is one step removed from action, it makes action more likely to succeed because each plan is better adapted for its particular purpose.

Intelligence thus contributes to adaptability in two ways: (a) By the construction of these schemas—not just one, but a large number, for all the different kinds of job that delta-one does. We can think of these schemas as mental models which embody selected features of the outside world. (b) By deriving from these schemas particular plans appropriate to different initial states and goal states. These plans can then form the basis of goal-directed action as already outlined.

These are two of the goals of delta-two. The first is a learning goal. The second goal is a planning goal: that of finding an appropriate plan for the achievement of a particular purpose. If such a plan is hard to find, even given the requisite knowledge, we call this activity *problem solving*.

## SCHEMA CONSTRUCTION

Knowledge structure of this kind cannot be communicated directly. They can only be constructed by activity of a learner's own delta-two, operating on his own delta-one. Many readers will recognise this as a constructivist position. But good teaching can greatly help intelligent learning. And indeed it must, if in ten or fifteen years we want children to acquire knowledge that it has taken centuries for the best minds of mankind to construct.

For this, we need to know something about how delta-two performs its first function, which is the construction of schemas. I use the word construction to mean both building and testing—doing something to the operand, and checking that this change takes it nearer the goal state. Suppose that we are constructing something in the physical world—a brick wall, or a transistor radio. In the case

TABLE 8.1
Schema Construction

| Schema Building | Schema Testing |
|---|---|
| **Mode 1** | **Mode 1** |
| from our own encounters with the physical world: *experience.* | against expectations of physical events: *experiment.* |
| **Mode 2** | **Mode 2** |
| from the schemas of others: *communication.* | Comparison with the schemas of others: *discussion.* |
| **Mode 3** | **Mode 3** |
| from within, by formation of higher order concepts: by extrapolation, imagination, intuition: *creativity.* | comparison with one's own existing knowledge and beliefs: *internal consistency.* |

of the wall, building and testing alternate. When we have put a few more bricks on the wall, we test for alignment and verticality. In the case of the transistor radio, we cannot test with certainty until it is built. These are delta-one activities; and we can see that building corresponds to those parts of the diagram marked PLAN and ACTION, while testing corresponds to the parts marked INFORMA-TION and COMPARATOR.

Now we need to transpose this up a level, and consider how delta-two performs its functions. The first of these is to provide delta-two with the schemas which are the basis for its adaptability, and successful action.

This kind of construction also involves building and testing. In the functioning of delta-two, we can distinguish three major categories of building, and three of testing. These I call *mode 1, mode 2,* and *mode 3* (see Table 8.1).

Although we can see connections between corresponding modes of building and testing, any mode of building can be used with any mode of testing; and we can use these more than one at a time. They are more powerful when used in combination.

The Doppler effect provides a good example of all three modes. Probably most of us have experienced the change in pitch of some fast-moving body as it goes past us—a car or motorcycle, a train, an aircraft passing overhead. We observe certain regularities, e.g., that it happens for all fast-moving objects that emit sound; that the pitch is higher when approaching then when receding. This is mode 1 building: experience.

By verbal communication and/or reading (mode 2 building), we may learn about the wave-theory of sound. This provides a general schema, with the help of which, either by our own delta-two activity or by further communication from others, we can explain this effect. We can now test our mental model of the Doppler effect by mode 1. To do this, we first make certain *inferences* based on the model: e.g., that the change in pitch will be greater when the object is moving faster. With the help of some maths, we can go further and calculate what this change in pitch will be for a given velocity; and thereby find a way of calculating the velocity from the change in pitch. These predictions can now be tested by experiment (mode 1). Success of these predictions will confirm our mathematical model, and perhaps encourage us to extrapolate it. By its independence of a particular context, this mathematical model lends itself well to extrapolation (mode 3 building). By this means the model that began as an awareness of certain auditory phenomena, and was mathematised for sound vibrations, is now applied to the kind of vibrations by which we account for the behaviour of light. And we can now use the same mathematics for calculating the velocities with which distant galaxies are moving toward or away from us.

I have taken a scientific example, one that I find impressive, to show the power of intelligence at a fairly advanced level. But we are using our intelligence almost from the day we are born, to build up the schemas by which we direct our actions at an everyday level. We continue to use it throughout our lives, to construct and keep up-to-date the schemas needed for the various ways in which we earn our livings—and also, to enjoy life when we have taken care of the material necessities. Here, I would like to correct a possible impression that my model is concerned only with the uses of intelligence for survival. I do suggest that our species evolved intelligence because of its advantages in the competition for survival. And it is still useful for that. But we can now use it for artistic and creative purposes too, in the same way as our prehensile hands may have evolved originally because of their advantages in climbing among trees, but they are useful now for playing the piano and painting pictures.

## USES OF OUR SCHEMAS

Our schemas have many uses, which fall into three main groups.

*Group 1.* For integrating knowledge, and making possible understanding. The resulting schemas provide a rich source from which delta-two can make predictions about events in the physical universe, and devise a wide variety of plans of action for controlling these.
*Group 2.* To help us to co-operate with others in a wide variety of ways.
*Group 3.* As agents of their own growth.

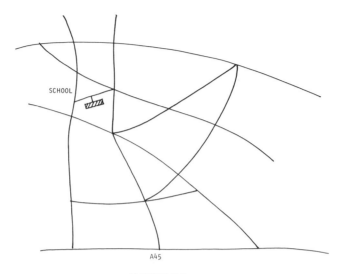

SCHOOL

P
•

A45

FIGURE 8.4

Group 1. The second of these has already been discussed in the present chapter; and the first was discussed in Chapter 3, where I wrote ''To understand something means to assimilate it into an appropriate schema'' (page 29). Here, I would like to show the connection between the two.

As we have seen, our schemas are the sources from which delta-two constructs plans for use by delta-one. It is by putting these plans into action that we achieve our goals, and these related abilities are important to us. But suppose now that we encounter something which is unconnected with any of our available schemas (see Fig. 8.4).

Delta-two cannot make any plan that includes this point (meaning, what it represents) either as an initial state, a goal state, or on the path between them. If the diagram is thought of as a map, then we are literally lost at this location, for it is not on our map. If metaphorically it represents a cognitive map, or abstractly it represents an unspecified schema, we are mentally lost. The metaphor is a very close one: we do not know what to do in order to achieve our goal. And this, in general terms, is our state of mind when we are confronted with some object, experience, situation, or idea that we do not understand.

The achievement of understanding makes connections with an existing schema (see Fig. 8.5). The new state has been brought within the domain of a director system. We now know that in such a state, we can cope. Metaphorically, and in some cases literally, we know where we are, so we can find a way to where we want to be. This change of mental state gives us a degree of control over the situation that we did not have before, and is signalled emotionally by a change from insecurity to confidence.

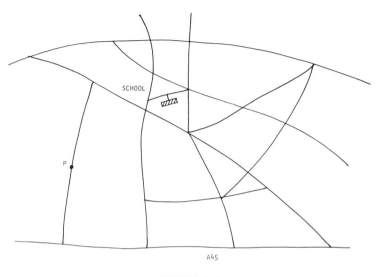

FIGURE 8.5

This explains the great subjective importance that many of us have long intuitively attached to understanding, and provides it with a theoretical explanation and justification. Understanding is at the centre of the processes by which intelligence enables us to survive, master our surroundings,[2] and create.

Group 2. In terms of the new model, co-operation between two or more persons requires that their various delta-one systems work together in some way, either to achieve a common goal, or compatible goals. (A simple example of co-operation for a common goal: two persons work together to move a heavy object, e.g., one puts rollers underneath while the other pushes. Of compatible goals: one person cooks a meal, everyone in the family eats it.)

So co-operation requires that the delta-one system of the people co-operating are using *complementary plans;* that is, plans that fit together to make an effective plan for the combination of two director systems working together. The successful working of any society depends on many and various people co-operating in many different ways for a great diversity of tasks, and this in turn depends on the availability of some way of arriving at complementary plans as and when required. The most successful way of doing this is by the existence of widely *shared schemas.* If individual plans are all derived from the same shared schema, this is a great step towards fitting them together. Co-operation also requires exchange of information, which is to say a *shared language* for the

---

[2]It has to be said that we have yet to understand, and master, the darker characteristics of *homo sapiens.*

schema, its concepts and their relationships. (If two people want to arrange a meeting, this will be difficult to plan unless their maps agree, they use the same calendar, they have the same way of telling the time, and speak a common language.) These shared schemas are also an important basis for co-operation within a profession.

Group 3.  The importance of our schemas as tools for future learning was also emphasised in Chapter 3 (pages 33ff.). Experiences that fit easily into our existing schemas are more readily learned, and better remembered, than those that do not. They also sensitise us to experiences that we would otherwise have ignored. Our schemas thus act selectively, enlarging themselves rather than other schemas which, although these too might be useful to us, we have not developed. This is one of the prices we pay for their benefits.

More recently, I have come to think of these properties of schemas as being organic in quality. They begin like seeds, that put out roots for gathering nourishment from the soil, and stems and leaves for gathering energy from the sun: in both cases, selectively. By making these into their own substance, they enlarge their roots and leaf structures, and thereby increase what they can take in. Schemas do this, turning experiences into knowledge.

Schemas have another property that is not included in the aforementioned analogy: that of creativity. They can grow from within, without any contribution from outside. Simply by combining or extrapolating ideas we already have, we can produce new ideas. Those that are subsequently given physical embodiments become important features of our culture and environment, from electric light to echo sounders and radar, from telephones to television, from microscopes to radio telescopes—with the help of an encyclopaedia, we can make a very long list. Some of these objects become, in their turn, extensions of our senses, which make possible still further growth of our schemas from outside sources. We also create mental objects which become tools for thought, and of these mathematics is one of the most versatile. This use of our intelligence is considered further in Chapter 11.

In this chapter, we have seen how the idea of a schema helps us to understand how our intelligence functions. In the next chapter, we shall consider in more detail the nature of schemas themselves.

# 9

## From Theory Into Action: Knowledge, Plans, and Skills

In this chapter, we think about how the knowledge embodied in a schema may be translated into successful action; and about how the adaptability which, in the schema, exists only in potential, is made effective. The first stage is the possession of an appropriate schema, which we may think of as *knowledge that*. The second stage is deriving from this an appropriate plan of action, which we may call *knowledge how*. The third stage is translating this plan into action, which we may think of as *being able*. Although different, these three are closely related, and it is their relations that we explore in terms of the present model.

### KNOWLEDGE

*Knowledge* here means organised knowledge, not a collection of isolated facts. (This is in sharp contradiction to what seems to be regarded as knowledge by the organisers of many television quiz programmes.) The diagram introduced in Chapter 8, and repeated here (see Fig. 9.1), is a useful way of representing a knowledge structure because it can be interpreted at three levels of abstraction.

1. As a road map. Here, each point represents a physical location.
2. As a cognitive map. Here, each dot represents a concept, and each line represents a connection between concepts.
3. As a generalized schema, representing an unspecified knowledge structure, and used to represent what these have in common. The terms *knowl-*

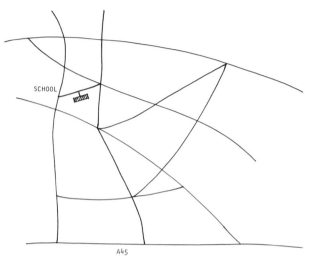

FIGURE 9.1

*edge structure, conceptual structure* and *schema* are used more or less interchangeably, depending on which aspect is being emphasised.

## WHY KNOWLEDGE MUST BE CONCEPTUAL

Next, I would like to look at the nature of these cognitive maps in greater detail. We can think of them as mental models derived from certain features of the outside world.

But as Heraclutus has told us, ''We cannot step twice into the same river.'' The present experiences from which (sometimes) we learn become part of our past, and will never again be encountered in exactly the same form. But the situations in which we need to apply what we have learned lie in the future as it becomes the present, or as by anticipation we bring it into our present thinking. It follows that if our mental models are to be of any use to us, they must represent, not singletons from among the infinite variety of actual events, but common properties of past experiences which we are able to recognise on future occasions.

A mental representation of these common properties is how, for many years now, I have described a concept; and for this process of concept formation I use the term *abstraction*. Concepts represent, not isolated experiences, but regularities abstracted from these. It is only because, and to the extent that, our environment is orderly and not capricious that learning of any kind is possible. A major feature of intelligent learning is the discovery of these regularities, and the organising of them into conceptual structures that are themselves orderly.

These conceptual structures, or schemas, are like cognitive maps only more so. We could think of them as cognitive *atlases,* of a rather special kind. If I want to drive from Coventry to Bristol, I use first a road map of England, on which Bristol appears just as a dot; then a street map of Bristol, in which my destination (say, the university) now appears as a dot; and finally, I might use a plan of the university campus that shows where I may park my car, the building and the room I want to find. If all these maps were in the same scale and detail, they would be unusable, for two reasons. They would be either too large to handle, or too small to show enough detail. And they would contain either too much, or too little, information.

These paper maps are symbols, which readily evoke the mental maps from which I derive my plans of action. They come separately. But in this case Bristol is also my birthplace, so I already have all these maps in my memory. Here, they are stored differently from the symbolic maps, in that they appear to be *nested* within each other. In my cognitive (i.e., mental) road map, it is as though there were a large dot representing Bristol. This is how I think of it while driving down the motorway. But as I turn down the slip road into the outskirts of Bristol, this dot expands into another map, a street map of the city. On this, in my thinking, the university now appears as a dot. Having parked my car and continuing on foot into the building, I expand this dot into a three-dimensional plan of the building.

The way in which I successively access these mental maps is less like getting out a different map and turning the pages than it is like looking at increasingly small areas of the same map under increasing magnification. So to describe this, I have used a metaphor from photography, in which we can buy a single lens of variable focal length. Looking at the same landscape, we can use this to give a wide angle view that we see in less detail, or by increasing the focal length we can get a larger, more detailed image of a smaller area. The detail has to be there: if we look at a dot on a road map under a magnifier, we do not see a street map of the city. And in my mental maps of England, there are towns that for me remain dots: I cannot access any further detail, beyond their overall location. But in nature, there is always more detail to be seen: city, street, building, bricks (or whatever), granular structure, molecules. So this kind of model is a good representation of nature.

To summarise these two complementary ideas, we may say that delta-two has a *vari-focal* ability to examine the contents of delta-one; and that in delta-one, knowledge is organised in schemas, now thought of as conceptual structures in which many of the concepts have *interiority.*

This provides a very economical and powerful way of storing knowledge. One of the desirable features of a mental model is to simplify the unthinkable complexity of the universe in which we live, enough to enable us to comprehend, think, and plan. But not to over-simplify. An ideal model would include all necessary information for a particular purpose, and nothing that was not needed.

In our schemas, as described above, we store all the detail we need for a wide variety of purposes, and use vari-focal access to scan them in the right amount of detail for the job in hand. This is one of the features of our intelligence which gives it such adaptability.

In mathematics, the process of successive abstraction that leads to the formation of concepts of successively higher order and generality offers a particularly strong example of the foregoing. It is an interesting exercise to go through some familiar process and analyse the levels at which one is working. For example, one is solving an equation. Already one has selected the right schema, mathematics, not cooking, nor reading music. Within this very wide-angle view one identifies, without detail, that the location we need is that of algebra. A closer look brings more detail into conscious use—an equation. If this is not in standard form, we need to examine it in closer detail before identifying it as a quadratic. Relating this now to our overall schema for quadratics, we have several well-practised routines for solving these. However, we are not yet at the starting point of any of these. To get the equation from its present state to one of these as a sub-goal state, we alternate between our quadratic schema and our more general algebraic schema, putting together a plan which is partly specific for this equation (to take care of this we need to move in and look at the specific details) and partly derived from general algebraic properties (for which these details are unnecessary). Sometimes we are concentrating on a single detail (e.g., possible factors of $-18$). All the time we are changing our focus, between the general and the particular, so rapidly that we do not need to think about it. That is, for our own purposes: we are good at quadratics. But to help those for whom these processes are not smooth and easy, we need to be more reflectively aware of what is going on.

## A RESONANCE MODEL

We all have enormous stores of conceptualised knowledge, collected together to form a large number of different schemas. Most of this is for most of the time quiescent, like books on a shelf, not books that are open and being read. Sometimes, to continue the analogy, we scan the shelves for a book on a particular subject; but often, it is as though the right book came to our hands of its own accord, open at the page we need. In the language of the present model, most of our schemas are inactive at any given time; and unless this were so, our consciousness would be overloaded with information, most of it irrelevant to our present needs. But for each new situation we encounter, usually an appropriate schema is activated, and within this schema, the relevant concepts. How do we do this? And how is it that the word ''group'' has one meaning for me when I am thinking about algebra, and a different meaning when I am thinking about classroom organisation?

To try to answer these and other questions, I have elsewhere (Skemp, 1979a) described a resonance model for the storage and retrieval of conceptualised information. It is by resonance that our radio or television sets becomes sensitive selectively, to a particular frequency from among the many electro-magnetic carrier waves that reach its antenna. It is by resonance that our ears distinguish between, and identify, sounds of different pitch and timbre. From starting points such as these, I came to think that resonance might offer a model for the activation (selective retrieval into consciousness) of schemas and concepts. Further developments of this model have suggested possible explanations of other features of concepts, such as the ways in which they may creatively interact to form new concepts, without further input from outside sources; and the phenomenon of perceptual distractors, which has been interestingly discussed by Behr & Post (Behr & Post, 1981).

## ASSOCIATIVE LINKS AND CONCEPTUAL LINKS

The difference between a schema and a set of isolated concepts is that in the schema, the concepts are connected to each other. The difference between understanding and not understanding a concept or a symbol lies in whether it is connected to an appropriate schema, or not. If the presence or absence of connections is so important, this suggests that the nature and quality of the connections themselves also deserve our consideration. In this section, we distinguish between two kinds of connection, which I call *associative* and *conceptual;* for short A-links and C-links; and we look at some of the qualities of these.

Figure 9.2 gives some examples of associative connections. If we look at the interior of each concept, the links here are also A-links. For example,

$$6\text{———}2\text{———}7\text{———}5\text{———}9.$$

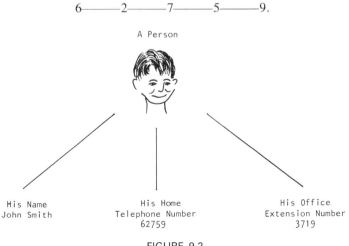

A Person

His Name
John Smith

His Home
Telephone Number
62759

His Office
Extension Number
3719

FIGURE 9.2

The only way we have of forming these connections is by rote memorizing. There is no regularity or pattern that would make possible intelligent learning. In contrast, here is an example of a conceptual connection:

3———6———9———12———15.

All the connections just represented by lines have something in common. We might, indeed, regard it as the same connection applied to different pairs of numbers, as shown in fig. 9.3.

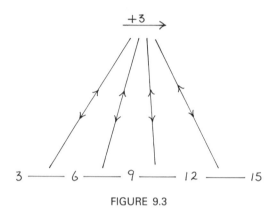

FIGURE 9.3

The difference in ease of learning is enormous. Imagine, on the one hand, having to learn a hundred digits arranged as twenty 5-figure telephone numbers; and on the other hand, having to learn the sequence printed above, to a hundred terms. (There are thus more than a hundred digits involved.) In the second case, we would not even need to learn all the numbers. We would learn a miniature knowledge structure, or schema, from which we can generate all the individual terms. Knowledge of the second kind is also more adaptable. If asked, we could say what is the 100th term in the series, or 99th, or indeed the 10th, 20th, 21st, 19th, and so on.

Different combinations of associative and conceptual learning are required for different tasks. Even in the telephone number example, we have a concept of a telephone number, what it is and what we can use it for. This represents what all telephone numbers have in common. If we make an international telephone call, there is a conceptual connection between the international dialling code, and the adapted form of the internal dialling code. In the learning, say, of English spelling, there is both associative learning as well as conceptual learning: associative learning where the spelling is irregular, and conceptual where it is regular.

In mathematics too, some of the connections to be formed are associative, for example the connection between a number concept and its symbol. But the great majority of the connections are conceptual. If, as happens all too often, associative (rote) learning is used, there is a great loss of efficiency and increase of labour involved. There is also a loss in adaptability, as already described. So intelligent learning requires that conceptual rather than associative connections are formed whenever this is practical.

As was mentioned in Chapter 8, the possession of a concept sensitizes one to recognise further examples of this concept. This applies likewise to conceptual connections, which are themselves a particular kind of concept. Thus, someone who has perceived the C-links in

$$3\text{———}6\text{———}9\text{———}12\text{———}15$$

is more likely to recognize them in

$$5\text{———}18\text{———}31\text{———}44\text{———}57,$$

even though it is not the same C-link as before.

Here is a third sequence:

$$1\text{———}6\text{———}11\text{———}16\text{———}21.$$

These three sequences have C-links that are both alike and different. They are alike in that there is a common difference between adjacent numbers, but they are different in that the amount of this difference is not the same. Thus, the conceptual quality of C-links gives rise to higher order concepts; in this case, the concept of an arithmetic progression. These can be used as a basis for further invention, such as

$$1\text{———}2\text{———}4\text{———}7\text{———}11\text{———}16.$$

Another difference between C-links and A-links is that once the C-link has been formed, there is something on which we can reflect. It becomes accessible to consciousness; we can put a name to it, and communicate it. "There is a common difference between successive terms. In the second sequence, this is 13." But when it comes to an A-link, it seems impossible to get beyond knowing that it exists. Here is yet another way in which the rote learning of mathematics handicaps the user.

## OTHER NAMES FOR SCHEMAS

The terms *schema* and *conceptual structure* I use interchangeably; the shorter one for convenience, and the longer one to emphasise two of its qualities, that its components are concepts, and that these are not isolated but integrated.

Concept Maps. We have also become aware, in Chapters 2 and 3 with a reminder in the present chapter, of the hierarchic nature of some knowledge structures, and particularly of mathematics. The mental map we have of our neighbourhood is not hierarchic. The different concepts involved are of about the same order, and we do not need to have the concept of the post office in order to form that of the local school. Likewise, we can learn the geography of Denmark without having learned that of Italy. In other areas of knowledge, however, certain concepts are prerequisite for the formation of other concepts. I remember this to my cost, when many years ago I was studying for a psychology degree and had also to pass exams in physiology. For this I had not the prerequisite knowledge of biochemistry, and passed my exams only by rote-memorizing. This is not an experience I wish to repeat; but it is one which is imposed on many children in their learning of mathematics, by the way in which they are taught.

In order to try and ensure that children do have available the necessary schemas by which they may understand what is currently to be learned, a useful device is a particular kind of schema that I first called a *concept dependency network*. Lately, I have come to use the more convenient term *concept map*, although the word ''map'' does not imply the property of order that is also important here. Once we have a schema well constructed, we can move around it in any direction we choose. But in the process of construction, we have to move in the direction from lower to higher order concepts. Applying this principle in detail to a particular area of mathematics, we get a partially ordered schema, showing which concepts are required for the understanding of others. The preparation of a concept map is closely related to the conceptual analysis described in Chapter 2, page 18. An example is in Fig. 9.4.

A concept map of this kind can be used in at least two ways: for planning a teaching sequence, and for diagnosis.

In the next section, we discuss the relationship between schemas and plans of action. Concept maps are a particular kind of schema, from which we can devise particular kinds of plan, namely teaching plans. They also give adaptability in teaching approach. The numbers on the concept map just shown suggest an order, but not the only possible order of approach. When building a three-story house, it doesn't matter much whether we build the front or the back wall, or the side walls, of the ground floor first; but all of these need to be there before the next floor is built. Similarly, in the learning of a particular mathematical structure, there may be several valid orders of approach. All, however, need to be based on the same principle, namely that to understand a new concept, learners must have formed and consolidated the earlier ones that are shown by the diagram to be prerequisite.

They are also helpful diagnostically. If a learner has difficulty with a particular concept, reference to an appropriate concept map may suggest that the roots of the problem lie further back, and indicates which areas we should check. For example, a child was having difficulty with problems involving comparison of two numbers, such as ''Wanda has four cookies, Sally has seven. How many more cookies has Sally?'' It turned out that this child had been taught to think of subtraction only as 'take-aways,' using only this name for the concept. In the word problem just given, there is no taking away: hence the difficulty. Reference

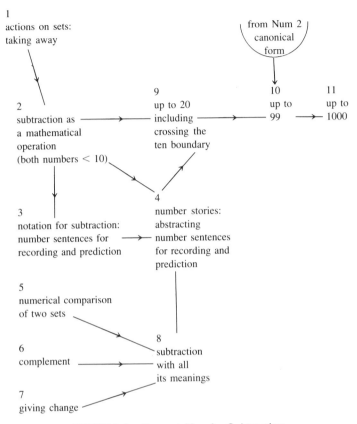

FIGURE 9.4.   Concept Map for Subtraction

to the concept map shows that the mathematical operation of subtraction is a higher order concept that subsumes what no less than four lower order concepts have in common. Each of these four, in turn, acts as a model for a different class of physical events.

Over the last few years, the construction of a number of these concept maps, for the first seven years of schooling, has made me aware of the conceptual complexity of these early stages of mathematics. Failure to take account of this is, I believe, one of the reasons why children continue to have difficulties.

**Frames and Schemas.**   The terms *frame* and *slot* (or *variable*) are used by Davis (1984), and other writers, with meanings which correspond closely to the meanings of the terms *schema* and *concept*. Here is an example (Davis, 1984):

> *Frames.* Perhaps the most interesting and the most provocative phenomena in information processing relate to the ways in which certain complex and highly interrelated bodies of information, often of an active type, can be represented in

memory. To deal with such matters, it has been necessary to postulate a very special kind of knowledge representation structure, known as a *frame*. (p. 45)

This is next discussed in the context of reading comprehension; but as Davis points out, the idea is quite general.

(i) a written sentence or paragraph cues the retrieval of one or more *frames* from memory; (ii) these frames then pose certain questions; (iii) the reader seeks answers to these questions, and inserts these answers into 'slots' (or 'variables') in the frame, thus bringing together *general* frame information and *specific information from this individual input;* (iv) where answers are not forthcoming from input data, the frame may make default evaluations; (v) if neither input data nor default evaluations are available to fill a key 'slot', the frame may refuse to function; (vi) an evaluation is made, to determine whether frame selection has been correct, and whether the variable slots have been filled correctly; (vii) from this point on, nearly all subsequent information processing will be based on data in this instantiated frame—and the 'primitive' data input will be ignored. (Davis, 1984, p. 47ff)

Here is a translation into the language of schemas and concepts.

A conceptual structure is called a *schema*. Among the new functions which a schema has, beyond the separate properties of its individual concepts, are the following: it integrates existing knowledge, it acts as a tool for future learning, and it makes possible understanding. (Chapter 3, this volume)

In the context of reading comprehension,

(i) A written sentence or paragraph activates one or more of the reader's available *schemas;* (ii) [no correspondence: as I read the description (Davis, 1984, p. 45), it is not the frame which poses the questions] (iii) the reader assimilates these words to his schema, and thereby understands the meaning of the sentence or paragraph; (iv) information which is not derived in this way may be provided by the rest of the schema, which embodies regularities (general features) of situations of this kinds; (v) if essential information is not available either from the written input nor from the rest of the schema, the person may not understand, and be unable to answer questions or otherwise act appropriately; (vi) an intuitive or reflective evaluation is made, to determine whether an appropriate schema is in use, and whether the sentence has been correctly understood; (vii) from this time on, nearly all subsequent thinking, planning, communication, and action, will be based on the total information provided by the combination of the particular instance and the general schema to which it has been assimilated.

The term *schema* has been in use for much longer, having been introduced into psychology by Bartlett (1932). And to me, the term *frame* is less suggestive of many the qualities of a schema, particularly its organic quality and the interiority of its concepts. Slots, or variables, do not make the important distinction between primary and secondary concepts, nor explain the process of abstraction by which we form progressively higher order concepts. Also, I am cautious in accepting as explanations of human thinking metaphors from what is conveniently, but misleadingly, called *information technology*. Computers process symbols, not information. They work at the level of syntax only, not semantics. My caution has deepened into mistrust since reading Weizenbaum's (1976) penetrating analysis.

However, I do not think it matters too much that we are using different terminology, provided that this does not prevent realisation that we are using the same ideas, or hinder communication. What matters most is to understand the importance of the ways in which we bring our organised knowledge into use for each new situation; and in this and other ways, Davis' book has made a valuable contribution. A similar theme runs throughout the work of Ausubel (Ausubel & Robinson, 1969):

> First, by *nonarbitrarily* relating potentially meaningful material to relevant established items in his cognitive structure, the learner is able to effectively exploit his existing knowledge as an ideational and organizational matrix for the incorporation, understanding, and fixation of large bodies of new ideas. It is the very nonarbitrariness of this process than enables him to use his previously acquired knowledge as a veritable touchstone, for internalizing and making understandable vast quantities of new meanings, concepts, and propositions, with relatively little effort and few repetitions. (p. 57)

## PLANS

By *plan* I mean a plan of action, and not a plan in the sense of a diagram. And we shall be thinking as much about mental actions, such as are involved in mathematics, as about physical actions, as are involved in driving a car. With this meaning, a plan is what we have to do, physically or mentally, to take the operand (whatever is being acted on) from the present state to the goal state.

As we saw in Chapter 8, from a map of a region we can make many different plans for driving from many different starting points to many different destinations. And we can make several plans for getting from a particular starting point to a desired destination, choose the best, and put this one into action. The operand in the previous example is the car, with ourselves and perhaps some passengers inside it. For solving an equation, we need a different kind of plan, in this case for mental action. Our starting place is the equation, as we see it written on paper. Our goal state is a mental state, that of knowing what is its truth set (i.e., the set of values of the variable that will make this equation a true statement). Other goals may also be involved, such as writing down symbols showing the steps in our thinking by which we arrived at this result. The goal state involved now is a state of mind of whoever reads these symbols, and the plan for achieving this goal state involves communication. More specifically, a pupil may need to convince a teacher that he has worked this out for himself; or one academic may wish to demonstrate to another the logical processes of inference involved, with a view to convincing the other of the validity of his work. The last reequires no less than three kinds of understanding, which are described in Chapters 12, 15, and 13. These are *relational understanding, symbolic understanding,* and *logical understanding.* Three kinds of schema are thus involved: for finding a solution and understanding what we are doing we need a *relational schema;* for communicating how we reach this solution, an appropriate *symbol system;* and for constructing a logical proof that our solution is correct, a *logical schema* in which the connections are inferences that if a particular statement is true, then that which follows must also be true.


## SKILLS: BEING ABLE


To have a plan of action is not the same as being able to put this plan into action. The passenger of a car may be well able to act as navigator, planning a route and telling the driver where to turn left, or right, or go straight on. But this passenger might nevertheless be unable to drive the car himself. It is the combination of having a plan, and being able to put it into action, which we call having a *skill.*

In general, schemas are not the only sources of the plans that can form the basis of skills. Some of these are genetically programmed; others are learned as habits. In both of these kinds, plan and action are fused, and the cognitive element is small. In both cases, the skills may be useful and effective in a particular set of conditions; but they lack adaptability. Some genetically determined plans of action (instincts for short) are so complex and effective that it is hard to see them as different from those that are the product of intelligence. But the difference is in their adaptability, as illustrated by the sad little story, given by Erikson (1950), of the homesick Englishmen who imported swallows from England to New Zealand. As Erikson writes, ''when winter came they all flew south and never returned, for their instincts pointed southward, not warmward'' (p. 88).

Examples of the lack of adaptability of rote-learned mathematics are found in Chaper 12. And even if we are (as we should be) concerned almost entirely with cognitively based skills, to have a plan does not guarantee our ability to put it into action efficiently and reliably. For example, a pupil may know exactly what to do in order to evaluate the product $(a - 17)$ $(a + 23)$, but still get a wrong answer due to an arithmetical slip, or a mistake in sign. The combination of having a plan and being able to translate this smoothly and reliably into action is what, in the present context, we mean by a *skill.*

At the level of action, it is hard to distinguish between a well-drilled habit, and a cognitively based skill of the kind described here. The difference, to say this yet again, is in adaptability. A child who has his multiplication tables as a cognitive skill will, if he has forgotten the result for $7 \times 9$, be able to reconstruct it, for example, by knowing that it is one 7 less than $7 \times 10$. In this knowledge, he also has a contributory example for use later on in forming an important higher order concept, that of the distributive property of multiplication over addition and subtraction. A child who is only well drilled in multiplication facts has neither of these advantages.

Let us not underestimate the advantage of having these facts readily, accurately, and reliably available in the form of well-practiced routines. But neither let us forget the importance of understanding, without which these routines become rote-learned habits, of little or no adaptability. To distinguish between skills in which the cognitive support is present or absent, I used in Chapter 5 the terms *automatic* (skill with understanding) and *mechanical* (a rote-learned habit). For the former, the term *fluent* is also a good one.

For successful performance, the best combination is a well-connected schema, with as many C-links as possible; and related to this, to be fluent in a number of routines for dealing with situations that are frequently encountered. These routines have a double advantage. They are an important contribution to our skills. And by reducing the amount of our conscious attention that we need to give to the routine aspects of the situation, we free attention for the novel, problematic aspects.[1] So this combination is particularly important as a foundation for successful problem-solving.

---

[1]The relationship between consciousness and novelty is an important one that I discuss further in *Intelligence, Learning, and Action* (1979) sections 2.7 to 2.9.

# Type 1 Theories and Type 2 Theories: from Behaviorism to Constructivism

The schemas described in Chapter 8 are of one particular kind among many.[1] They are mental models embodying selected features of the outside world. Even at a common-sense level, they greatly increase our power to understand, predict, and control events in the physical world. These common-sense schemas represent for the most part objects and events directly accessible to our senses. We use them for our everyday living in much the same ways as mankind has done before us for thousands of years. But the word processor on which I am writing this chapter cannot be understood at a common-sense level. It is a product of manufacturing techniques derived from mental models of objects and events far beyond the reach of our senses. Yesterday evening, on a similar cathode ray tube in our living room, we watched pictures of the planet Uranus and its moons, transmitted back to Earth over many millions of miles by the space craft Voyager 2. Mental models of this kind are a relatively recent arrival in our history. The uses to which we put them fall into the same groups as have already been described in Chapter 8, pages 111ff.: but their power and sophistication enable us to attain goals that would have been thought impossible until recently.

We can use our mental models at a practical level, to invent objects that do not yet exist in the physical world, but might. It may then become one of our goals actually to make this object; and for this purpose, we would use our schemas in another way, for devising plans of action. This can happen at an everyday common-sense level, as when we make ourselves a new gadget of some kind; at

---

[1] Other kinds of schema include those from which are generated works of fiction, music, and choreography.

the level of a craftsman or technician, as when someone designs and makes an improved kind of wheelbarrow. Or it can result in bringing about marvels that not long before would hardly have been thought possible: intercontinental transmission of pictures via geostationary satellites, the insertion of pacemakers into tired hearts. Here we are talking about the applications of science in technology; and the starting points for these achievements of men's thinking are scientific theories.

The distinction between a theory and a common-sense mental model is one of degree rather than of kind: the degree of abstraction and generality. Thus, I would regard the visual, tactile, and kinaesthetic model of my bicycle that I first formed as a child as a common-sense model; that of the chain drive, which converts the slower rotations of my pedals into faster rotations of my rear wheel, as the first step toward a theory. If this were developed into a general model of velocity ratios, related to the ratio of the numbers of the cogs in the two wheels, I would call this a *theory*—a fairly simple one. Having noticed that a wheel by itself stayed upright as long as it was rolling along, I had a mental model that I would not yet call a theory. When at university I learned the mathematics of spinning tops and gyroscopes, I then had a theory. This one I found quite difficult.

In the present context, a theory is a mental model of a more abstract and general kind than those that we construct and use at an everyday, common-sense level. Sometimes, what we observe, and its causes, are at about the same level of abstraction. I see a tree in motion, and I can feel on my face the wind which causes this. At other times, this is not the case, I see an image on my television screen, but many of the causes of this are beyond what is observable, and beyond what can be understood at a common-sense level even if we could observe it. The more remote the causes of events are from what is directly accessible to our senses, the more we need the help of a good mental model; the more *abstract,* with the meaning of Chapter 2, it will be. Also, the harder such a model is to construct. I have discussed this aspect of theory construction more fully elsewhere (Skemp, 1979a). Here, I want to move in another direction, which is to distinguish between two types of theory, and to consider some of the consequences of this distinction when we come to theorising about the learning of mathematics. In Chapter 11, we consider the question that also arises: Which kind of theory is mathematics? This also looks like an important question: for how can we usefully think about teaching it if we do not know what kind of a theory it is that we are trying to teach?

## TYPE 1 THEORIES AND TYPE 2 THEORIES

A *type 1 theory* is an abstract, general, and well-tested mental model of regularities in the physical world. It embodies what are sometimes called *laws of nature,* and to qualify for this description it must have explanatory and predictive power. All the natural sciences such as chemistry, astronomy, metallurgy, aerodynamics, electromagnetic theory, genetics, are thus type 1 theories; and in their respective fields of application, theories of this kind have been remarkably successful.

The theory outlined in the preceding two chapters is, however, of a different kind from the one just mentioned because it is not a model of regularities in the physical environment. To accommodate this, we need to start a new category. When we have done so, we find that other theories also belong there.

A *type 2 theory* is a model of regularities in the ways in which type 1 theories are constructed, and by which plans of action (for execution by delta-one) are derived from these theories. It is a mental model of the mental-model-building process. Examples of type 2 theories are: Piagetian theory, van Hiele's theory about the learning of mathematics, and my own theory of intelligence. I also see the constructivist group of theories as belonging to this category.

We also need to examine the possibility—indeed, the likelihood—that the ways in which these two kinds of theory may appropriately be *used* are also different. Type 1 theories are concerned with operands in the physical world. Collectively, they form the natural sciences, and with their associated technologies these have given us great success in manipulating our physical environment. Type 2 theories, in contrast, are concerned with what happens in our own minds, and those of others. Their operands include type 1 theories, and whatever mental objects go to make up type 1 theories: concepts and relations between these, conceptual structures, statements, conjectures, hypotheses, and all those processes by which type 1 theories are built and tested. (And not only those processes—but I want to keep the discussion within reasonable bounds.) To think, or to assume unthinkingly, that it is appropriate to use type 2 theories in the same way as type 1 theories leads rather easily to a manipulative attitude towards other people. This, I believe, is another way in which behaviourism sometimes went wrong.

## THEORIES AND METHODOLOGIES

A *methodology* is a collection of methods for constructing (building and testing) theories, together with a rationale that decides whether or not a method is sound. This includes both constructing a new theory *ab initio,* and improving an existing theory by extending its domain or increasing its accuracy and completeness.

Methodology and theory are thus closely related, and the construction of a successful theory will largely depend on the use of appropriate methodology. It might therefore be expected that researchers would refer explicitly to this relationship. This is not usually the case, although there are notable exceptions. For example, Steffe (1977) writes:

Constructivism, an epistemological theory, has not yet produced a theory of mathematics learning. However, several principles central to constructivism have been used to provide powerful analogies for building models in the teaching and learning of the whole number system. The central purpose of this paper is to outline a continuation of the construction of such models using a methodology called the teaching experiment.

Ginsburg (1977) also states explicitly his theoretical position and methodology.

In the spirit of Piaget, I try to show how the child's mind operates and develops as he or she encounters mathematical problems in and out of school. . . . The primary method is the in-depth interview with children as they are in the process of grappling with various sorts of problems. (p. iii–iv)

Where a researcher has not explicitly indicated the grounds for his choice of methodology, there are several possible reasons.

1. It may be that all those whom he expects to read his report use the same theory with its associated methodology, which he takes for granted and does not seek to challenge. This is usually the case with researchers in the natural sciences, such as electricity and magnetism, chemistry, and atomic physics. Research of this kind falls into the category that Kuhn (1970) calls *normal science*. It is certainly not the case with mathematics educational research, nor with the psychological research often used by education researchers as their starting point. In both of these fields it is easy to identify a number of alternative theories, none of which is so universally accepted that it may be taken for granted that both writer and reader are using it.

2. It may be that a theory, or at least a general theoretical position, is clearly implied by the content of the report. For example the title of a paper by Allardice (1977), *The Development of Written Representations for Some Mathematical Concepts* makes it clear that the author takes the position of the cognitive psychologists, in which concepts and symbols are importantly different, rather than that of the behaviourists, to whom a concept is a common response to a class of stimuli (which is to equate a concept with its symbol).

3. The researcher may be at the stage of making systematic observations, not yet organized into a theory. Even so, a theoretical position, that is to say a category or kind of theory, is implicit in the kind of observations which we made and the conditions under which they were made. For example, written tests administered to groups of children imply one kind of theoretical stance, whereas naturalistic observation and individual in-depth interviews imply a different kind.

4. Often, however, it is difficult to avoid the conclusion that the researcher has used a particular method without due consideration in relation to a meth-

odology. By a *method* I mean what a researcher does, his plan of action; and by a *methodology* I mean the more general rationale from which he derives a particular method and by which he can justify it. A person who uses a method unrelated to a methodology is thus in somewhat the same position as a pupil who uses algorithms in mathematics without having the underlying mathematical schemas from which the algorithm is derived, and by which it can be understood as a correct procedure.

Both (2) and (3) are acceptable positions; (4) in my view is not. In the natural sciences, (1) also is acceptable; but not in the field with which we are at present concerned (although it might be so within certain groups, such as the members of a particular research group).

## TYPE 1 METHODOLOGIES
## AND TYPE 2 METHODOLOGIES

A *type 1 methodology* is concerned with constructing (building and testing) the models that delta-one requires for its successful functioning. When constructed, these models are type 1 theories.

A *type 2 methodology* is concerned with constructing (building and testing) models of how type 1 theories are constructed, and how particular plans of action are derived from these. When they have been constructed, these models are type 2 theories.

The great successes of the natural sciences has, in the past, led to the unthinking assumption that by the adoption of similar methodologies for the development of theories about how we think, learn, and relate to one another, equally sound and successful theories would be developed. If the distinction between these two types of theory is accepted, it follows that this assumption must now be carefully examined, since on the face of it, there is no reason why these two types of methodology should be the same. They do, however, have this in common: the construction of both type 1 and type 2 theories is an activity of our intelligence. So Table 8.1 on page 110, in which are summarised the ways in which mental models are constructed, offers a suitable starting point.

The left-hand column of Table 8.1 corresponds to changes of state of the schemas in delta-one; the right-hand column corresponds to activities of the comparator in delta-two, testing that these changes are for the better. And as was said in Chapter 8, these are more powerful when used in combination.

## METHODOLOGIES IN THE NATURAL SCIENCES

Each of the natural sciences has its distinct methodology. Different methodologies are used for astronomy and for aerodynamics, for electronics and for geophysics. These do, however, have certain overall principles in common, and together these constitute what is called *scientific method* (Popper, 1976). These accord well with the modes of schema construction summarised in Table 8.1.

For building theories in the natural sciences, all three modes of schema building are appropriate and accepted. There is no shortage of examples for each mode. Faraday's theory of electromagnetism was based on observations of deflections of a compass needle (mode 1 building). The apprenticeship of each new generation of scientists includes the study of the discoveries of the past (mode 2 building). From the conceptual structures built up in these ways, the mind by its own creativity generates new ideas. These may surface in a variety of ways, some of which might hardly be thought of as scientific: such as Kekule's famous dream of a snake eating its own tail, from which he gained insight into the structure of the benzine molecule. But other famous scientists (see Ghisellin, 1952) have also testified to the importance of intuition in their own discoveries. We need all the ideas we can get in arriving at new discoveries, and should not be too fussy about their sources.

What is essential is that these new ideas be properly tested before being accepted into the scientific body of knowledge. For this, all three modes are important: but in the natural sciences mode 1 is indispensible. The motto of the Royal Society, *Nemine in Verbo*, tells us that it is not by words alone, but by the test of the repeatable experiment that a (type 1) theory finally stands or falls.

This fits in with the present model. If the purpose of constructing type 1 theories is to increase the powers of delta-one relative to the physical world, the physical world is where they must prove their success. Other criteria, such as economy, coherence, intelligibility, are also important. They help to make a theory more useable, by facilitating the conversion of knowledge-that into knowledge-how.

## METHODOLOGIES USED
## IN MATHEMATICS EDUCATION RESEARCH

### Behaviourist and Neo-Behaviourist Methodologies

The dominant influence that this school exercised over many years in the past has now diminished, and relatively little of what is innovative in current research is behaviourist. However, this still provides a good example of the relationship between theory and methodology; and there are important lessons to be learned by analysing the errors which, with hindsight, we can see to be inherent in the behaviourist approach. If we do not learn from these, we are in danger of falling into errors of the same kind although in new disguises.

The growth of behaviourism is closely associated with the efforts of academic psychologists to establish psychology as an accepted science. It is very understandable that these efforts took as their model the natural sciences, which even in the early days of psychology were proving their power in enabling us to shape our physical environment; and since then have shown an exponential rate of growth.

Characteristic methods in all the physical sciences are:

1. the replicable experiment, by which others can verify the results of an individual researcher, as a precaution against experimental error and as a pre-requisite for the general acceptance of these results;
2. measurement in standard units, without which experimental conditions and results cannot be described accurately enough for the above;
3. the isolation and manipulation of independent variables, so that their separate effects on the dependent variables can be measured.
4. quantitative as well as qualitative statement of results.

To use these methods in experimental psychology (and subsequently in the application of this kind of psychology to educational research), adaptations were necessary. To take a simple example, an experiment in the electrolysis of a saline solution is replicable because two samples of NaCl, and two samples of pure water, are identical; and the electro-motive force and current can be measured by test instruments internationally standardisable with a high degree accuracy. But no two persons are identical; so it becomes necessary to work with groups of subjects, on the assumption that individual differences that affect the result of the experiment are random, and that their overall effect on the dependent variable, when averaged, is close to zero. Thus, although it is not expected that experiments will be replicable with single subjects, this is expected to be the case with comparable groups of subjects. This introduces the need for simple statistical treatment of the results. Beyond this, the separate manipulation of independent varibles is also sometimes hard to achieve with groups of human subjects; so instead, their effect is teased out from the set of measures which represent the outcome of the experiment by more sophisticated statistical techniques such as analysis of variance, factor analysis. Another procedure designed to ensure replicability is operational definition of the variables—that is, in terms of publicly observable behaviour of the experimenters and of their subjects.

To reject behaviourist models because they are mechanistic is understandable but, in my view, not a good reason. Carpenter (1979) points out that "the relevant question is pragmatic. Which model is more fruitful for adequately explaining and predicting behaviour?" (p. 6) And although teaching methods based (consciously or unconsciously) on behaviourist models have been remarkably successful in bringing about various kinds of habit learning such as bar-pressing by rats and kicking a ping-pong ball by pigeons, it is a hard fact that they have been remarkably unsuccessful in bringing about the higher forms of learning, in which man most differs from the laboratory rat and pigeon, and of which mathematics is a particularly clear example.

As well as the pragmatic objection to behaviourist models, that they haven't worked, there are other criticisms to be made, on the grounds that they make *category errors*.

In constructing psychological and educational models similar to those that have proved so successful in the natural sciences, an implicit assumption has been made that on examination appears questionable. This is, that the kinds of objects whose qualities we seek to discover, abstract, and embody in our models are the same in both cases: or in other words, that different though the objects themselves may be, these differences are not such that a different kind of model is required. To give an analogy; although English, Russian, and Greek, are written in different scripts, these all consist of a basic set of symbols from which are constructed words, the words then being put together to make sentences. So a person whose first language was English would not have to make any major change in his approach if he wanted to learn to write either of the others. Chinese and Japanese ways of writing, however, are based on different principles. Whereas in English, Russian, Greek, the separate letters represent sounds (albeit rather loosely), in Chinese and Japanese the characters represent meanings. This would have to be explained at the outset to a new student of these languages. If nobody told him, and he never found out for himself either, he would make little progress because of his error in thinking of the new kind of writing as if it were in the same category as those that he knew already.

The first of the category errors that I believe to be inherent in any behaviourist model is that whereas our physical environment is indifferent to our activities in shaping it, our fellow humans are not. Any attempt by A to shape the behaviour of B implies some degree of loss of freedom for B, whether this be realised or not. This raises the possibility (to put it at its least) that consciously or unconsciously, B will seek to preserve his freedom by resisting having his behaviour shaped. Whether or not he resists, and how much, will be likely to vary between individuals: and will depend partly on how each construes the situation, again not necessarily consciously. Where this factor exists, or there is a strong prima facie possibility of its existence, I suggest that to ignore it leads to a category error.

A second category error is made when symbols are equated with concepts: when a sound or a mark on paper is equated with its meaning. To a behaviourist, the utterances

$$a(x + y) = ax + ay$$

and

multiplication is distributive over addition

are different behaviours. But to a mathematician, these are just two among many different ways of expressing the same meaning; and to a maths educator, what is important is that the learner realises this, understands the meaning, and can show that he does by applying the schema to a variety of examples. So for researchers into mathematical education, the distinction between symbols and concepts is one that it is essential to preserve.

The third and most important category error that I believe to be characteristic of behaviourist models is that they fail to distinguish between type 1 theories and type 2 theories. This distinction has already been discussed.

## Diagnostic Interviews and Teaching Experiments

In strong contrast to the aforementioned, both in methodology and theory, is the work of Piaget and his adherents. A clear and concise account of Piaget's methodology, and its origins, is to be found in Opper (1977); from which the following extracts are taken.

> In the mid-1920's, at the start of his career, Piaget worked in Simon's psychological laboratory in Paris where one of his duties was to standardise a French version of a series of Burt's reasoning tests . . . While engaged in his work, Piaget became particularly interested in the incorrect responses given by the younger children and decided to carry out cognitive studies in order to discover the underlying reasons for incorrect answers in younger children and correct ones in older children. (p. 90)

> Since no adequate research method existed for the type of studies he wished to conduct, Piaget created his own. Familiarity with the clinical interviews[2] used in the medical field led him to design a similar method for the study of reasoning in children. . . . The essential character of the method is that it constitutes a hypothesis-testing situation, permitting the interviewer to infer rapidly a child's competence in a particular aspect of reasoning by means of observation of his performance at certain tasks. . . . For the most part the experiment involves both a concrete situation with objects placed in front of the child and a verbally presented problem related to this situation. . . . At the start of each session, the interviewer has a guiding hypothesis about the types of thinking that the child will engage in. . . . For each item the interviewer then asks a series of related questions which are aimed at leading the child to predict, observe, and explain the results of the manipulations performed on the concrete objects. It is these predictions, observations and explanations that provide useful information on the child's view of reality and his thought processes. (pp. 92–93)

> The interviewer then tests his original hypothesis on the basis of the child's verbal responses and actions. If further clarifications are required, he asks additional questions or introduces extra items. Each successive response of the child thus guides the interviewer in his formation of new hypotheses and consequently in his choice of the subsequent direction of the experiment. (p. 93)

---

[2]For me, the word *clinical* has mainly medical connotations. I prefer the word *diagnostic*, as conveying the present meaning better: dia- meaning *through*, and gnosis, *knowledge;* and hence, knowledge of what is beyond the observations.

The foregoing methodology may be contrasted, point by point, with the methods listed as characteristic of behaviourist methodology in the preceding section. Instead of a replicable experiment, we now have individual interviews, no two of which are exactly alike. Instead of experimental designs carefully planned in advance, and executed so far as possible according to these plans in every detail, we have experiments in which only the initial situation and hypothesis are prepared in advance, new hypotheses and procedures being successively introduced according to the results of the experiment thus far. Instead of the outcome being measured in standardised units, it is presented descriptively. Often extracts from the child's verbal responses are given verbatim, together with the experimenter's inference therefrom. Fourthly, in the behaviourist methodology the experimental results are usually given in the form of some kind of array of figures, such as a correlation matrix, table of means and standard deviations, analysis of variance, together with significance levels, from which conclusions are derived that the experimental hypothesis as stated at the outset if thereby confirmed or infirmed. In contrast, the outcome of a Piagetian experiment is presented in the form of some general statement giving a synthesis or overview of the final state of the experimenter's thinking, resulting from the successive modifications of his original hypothesis during the course of the experiment. And finally, the Piagetian approach is much more time-consuming, relative to the number of subjects from whom data is collected, than the behaviourist. The amount of experimenters' time that it takes is a major practical difficulty in Piagetian-style research.

What are the implicit assumptions underlying these sharply contrasted paradigms? Concentrating on those most directly relevant to the present volume, and over-simplifying for the sake of emphasis, I suggest that these may be summarised as follows:

Behaviourist Paradigm.   What we are interested in is subjects' publicly observable behaviour, and this is shaped by conditions external to the subjects. These conditions can be defined operationally, and controlled with a fair degree of precision by an experimenter or teacher. Factors internal to the subjects, and especially those particular to individuals, are random in their occurrence and can therefore be eliminated by appropriate statistical techniques.

Piagetian Paradigm.   What we are interested in is the mental processes that give rise to the subject's observable behaviour, and these are mainly the result of processes internal to the subject. These vary between different individuals, and between the same individual at different ages; and the differences are as important as the likenesses. To investigate these we need to work with individuals in a one-to-one relationship with the experimenter, making hypotheses about underlying mental processes which are tested against a variety of observable behaviours.

# TEACHING EXPERIMENTS

Classical Piagetian theory takes little account of the function of instruction. In the context of education, however, the relations between instruction and learning, together with the nature and quality of this learning, are among our chief areas of concern. So it is not surprising that some researchers have based their investigations on the methodology of the teaching experiment. Among these are the constructivists.

A summary of six principles of constructivism is given by Steffe, Richards, and von Glassersfeldt (1979). Among the key features, as there presented, are the following.

> Knowledge is viewed as pertaining to invariances in the living organism's experience rather than to entities, structures, and events in an indepently existing world. Mental operations are part of a total structure, and structure is seen in the organisation of operations. Different surface behaviours of a child may be interpreted as springing from the same cognitive structure. The structure of the learning environment must be considered within two frames of reference. On the one hand there are the operational systems controlling the child's experiences and, on the other, there is the content to be learnt. Concepts, structures, skills, or anything that is considered 'knowledge' cannot be conveyed ready-made from teacher to student or from sender to receiver. They have to be built up, piece by piece, out of elements which must be available to the subject. (p. 29–31)

Readers will not fail to notice the close correspondence between the ideas in the previous quote and those described in the present volume. When first reading the paper from which these extracts were quoted, I found myself in the position of Moliere's character in *Le Bourgois Gentilhomme,* who discovered that he had been speaking prose all his life without knowing it. In my own case, I have been a constructivist since the early 1960s (Skemp, 1962), although I did not meet this term until the mid 1970s. And it was not until more recently (Skemp, 1985) that I personally experienced the value of the teaching experiment.

This methodology may be regarded as an extension of that of the diagnostic interview, in which the purpose is to make and test hypotheses not only about the nature of a child's thinking at a particular time, but about how this thinking develops from one stage to another. It is summarised by Steffe (1977) as follows:

1. Daily teaching of small groups of children by the experimenters,
2. intensive observation of individual children as they engage in mathematical behaviour,
3. prolonged involvement with the same children over periods ranging from about six weeks to the academic year,
4. clinical interviews with children, and
5. detailed records of observations through audio-video taping and the written work of the children.

The foregoing accords very well with the present model. A major emphasis of the latter is that what we can learn with understanding depends on our currently available schemas. These schemas cannot be observed directly in children, or other learners: they have to be inferred from their responses. The kind of responses we need for this purpose, which includes distinguishing between what has been learned with understanding and what has just been memorised as a rule, are not written responses to standardised questions, but the kind that are obtained in the situation of the diagnostic interview. So the combination of a teaching situation with the diagnostic interview offers opportunities for inferences both about the states of children's schemas at various stages in their learning, and about the processes by which they progress from one stage to another.

This line of thinking also makes good sense in relation to education. Just as a major field in which the natural sciences prove their worth is in what they can help us to achieve in the physical world, so a major field in which type 2 theories must show their worth is that of education. For this, researches based on teaching experiments get off to a good start.

My own experiences in this field suggest that from an approach very like that described by Steffe, but less intensive, much of value can still be learned. From 1980 to 1985 a colleague and I were working for one morning a week in schools with children aged 5–11, our main purpose being the field testing and revision where necessary of teaching methods and materials based on the theory described in the present book. The position taken was that although there is always more that we need to know about how children learn mathematics, the time had now arrived when there would be more benefit to the children from translating into classroom practice the theory as at present developed than from attempts to polish the theory still further. This was still a teaching experiment, although the purpose was one of curriculum testing and development rather than theory construction; and I found the experience illuminating and rewarding.

The teaching situations were of three main categories: discussions led by the experimenter, activities by the children working together in pairs or small groups, and mathematical games for 2–6 players. Children's informal discussions with each other about what they were doing, and also the explanations they gave to each other, both as help and as justification (e.g., for a move they had made in a game) were very rewarding. Important also was their use of mathematically structured materials. Much of what we learned was from observations of this kind, and subsequent discussions with the teachers involved. Sometimes we also asked the children questions, such as "Can you explain how you knew that?"

## Observation I

The experimenter (Ainley, J., personal communication, 1985) was working with three five-year-olds. She asked them if they could tell her what 5 + 4 was. The first child used counters. He counted out five of these, then he counted out four more, and finally he counted them all. The second child used his fingers, using the method of counting on. The third child looked at the ceiling. All three children gave the same answer, nine. My colleague asked the third child: "I could see what the others did to get the answer, but I couldn't see what you did. Could you tell me?" The child replied "Five and five is ten. Four is one less than five, so the answer is one less: nine."

Interpretation. The first two children were both at the stage of schema building, using mode 1 (physical experience). The second had a more advanced model than the first because counting on implies at least an intuitive awareness that the first of the sets to be united is part of the set resulting from this union. The third had a relational schema, from which he was able to extrapolate from something he already knew (5 + 5 = 10) to construct a new result (5 + 4 = 9). The chain of inference is good for a five year old. This and many other observations have led me to view with admiration the level of thinking of which children are capable, if we allow them to preserve their natural abilities (Ginsberg, 1977).

An interesting follow-on from Observation I was provided by a teaching experiment several months later with different children, in a different school.

## Observation II

In this activity, I was concerned with developing fluent recall of addition facts. (Please remember the distinction between this and rote learning.) The learning situation was provided by a game for two. One player turned over cards on which were written all pairs of number additions, from 1 + 1 to 5 + 5. This player had to say the result if he could; the other had a linear slide rule, which he used either to check the result if there was doubt or to obtain it if necessary. (A linear slide rule is made from two number tracks side by side. It is a physical embodiment of counting on.) The teacher had given me two children whom he described as backward seven-year-olds.

The child with the cards had given the correct result for 5 + 5; and it chanced that the next card he turned over was 4 + 5. He did not know the answer to this. Instead of telling him, his partner said "Well you know 5 + 5, don't you? You've just said it. So what's 4 + 5?"

Interpretation. No so backward, the latter child! Not only did he have a relational schema, but he had an intuitive grasp of the difference between helping someone to construct his own result, and just telling him. Almost, a seven year old constructivist! When I related this to their teacher afterwards, he said "Oh yes, he can do things if he wants. But he just doesn't try." I restrained myself from saying that this might depend on whether he found the work interesting or not.

In the end, theory construction and curriculum development must be complemented by increased professionalism on the part of teachers themselves. It is therefore encouraging to report that our experience of working in this way over a number of years has also given rise to the development of a approach to school-based in-service education which shows good promise. In this, teachers develop their own theoretical understanding in close relation to their own experience and classroom needs, working with their own children in their own classrooms. This has been described elsewhere (Skemp, 1983).

# 11

# Mathematics as an Activity of our Intelligence

For many years, I have held the view that mathematics is a particularly concentrated example of the functioning of human intelligence, and that this is one of the reasons why it is so powerful. But until the new model of intelligence was developed, this was largely an intuitive, although strongly held, conviction. In the first part of this chapter, I show how mathematics can be seen as an important particular case of the general model described in Chapter 8, and in the second part, I consider a question that arises from Chapter 10: What kind of theory is mathematics?

In the new model two main functions of intelligence were distinguished: constructing schemas, and deriving plans of action from these. So in applying the model, we need to consider both the learning of mathematical structures, and also their use as a source of plans of action, using this to mean both mental and physical action. Our available modes for schema construction have already been discussed in general, and we return to this in the particular context of mathematics. But the process of moving toward any particular learning goal is influenced by our conception not only of the nature of this goal, but also of what we shall be able to do when we have achieved it that we could not do before. So here, I want to focus on the uses to which we can put our mathematical knowledge when we have acquired it.

## USES OF OUR MATHEMATICAL SCHEMAS

In Chapter 8 were listed three main categories for the many uses of our schemas.

*Group 1.* For integrating knowledge, and making possible understanding. The resulting schemas provide a rich source from which delta-two can make predictions about the physical universe, and devise a wide variety of plans of action for controlling these.

*Group 2.* To help us to co-operate with others in a great variety of ways.

*Group 3.* As agents of their own growth.

## Uses of Mathematics in Group 1

I have already emphasised the process of abstraction, by which we build mental models that can be used in a variety of situations. These models represent, not single events (they would be altogether useless if they did), but regularities among events. Mathematical schemas show particularly well the additional power that results from repeating this process of abstraction. First, we find regularities among our experiences of the physical world—among the various sensory inputs to delta-one. These regularities are embodied in what I have called *primary concepts.*

Then we find regularities among these regularities—we form secondary concepts. We continue thus, finding higher and higher order regularities. These are embodied in concepts that are increasingly abstract. Because this means that they are also increasingly remote from our direct sensory experiences of the outside world, it would be easy to think that they were not a lot of practical use. In fact, the opposite is the case. Following Dewey (1929), I do not tire of saying "There's nothing so practical as a good theory"; and this applies with full force to mathematics. Although the process of repeated abstraction might seem to be taking us away from the hard facts of the physical world, when this is done in the right ways (and mathematics is one of these) it takes us from the surface of events to their essence. A good theory enables us to penetrate beyond the observables to the heart of the matter.

The process of abstraction provides us with mental models whose application is not restricted to particular cases. This applies to all schemas; but mathematical schemas provide us with models that are so multi-purpose as to be remarkable when we begin to think about it.

The natural numbers themselves are concepts that are independent of the spatial distribution of a set of objects; and even of what these objects are, including all their properties except separateness.

When we have left out what the objects in a set are, where they are, and what they are like, you would not think there would be much left. In fact, we are left with the natural numbers. With the operations of addition and multiplication and their inverses, and the general concept of a unit, we have a basic "kit" from which particular models can be made to represent almost any physical quality or event: distance, time, velocity, mass, acceleration, angles, angular velocity and acceleration, linear momentum, angular momentum, rotational inertia, physical force, electromotive force, electrical resistance (to d.c.), electrical impedance (to a.c.), frequency,—this list, although long, could easily be continued. From among these, let us begin by considering velocity, which is the ratio of distance to time.

*Distance* is independent of the particular starting and finishing points in space, and of the object that is moving. It does not even have to be the movement of a physical object: It may be a wave front—a flash of light, a radio signal from a satellite, a sound wave from an echo sounder. *Time* is independent of particular starting and finishing points in time. So the model

$$v = \frac{d}{t}$$

will do for the movement of walkers, cyclists, motorists, aeroplanes, planets: of physical objects from snails to stars, and movements as slow as continental drift and as fast as the speed of light.

All this is so familiar to us that we easily take it for granted. But as long as we do so, we overlook one of the most remarkable features of mathematical models—their enormous generality and adaptability. Man is the most adaptable of all animals, by virtue of his intelligence. One of the chief things our intelligence enables us to do is to construct multi-purpose, adaptable models whereby we can direct our behaviour to achieve a wide variety of goal states in a wide variety of conditions. And mathematical models show this quality particularly clearly, as soon as we begin to look at them in this way.

The process of successive abstraction gives a schema that is multi-purpose in another way—it allows us to work at different levels of generality, using our vari-focal ability. Just as the single statement

$$(a + b)x = ax + bx$$

contains an infinity of particular statements like

$$(7 + 2) \times 3 = 7 \times 3 = 2 \times 3$$

so the electrical formula

$$W = IE$$

enables us to construct particular models for an infinity of particular cases.

For example a 120 volt 60 watt electric light bulb takes a current

$$
\begin{aligned}
I &= \frac{W}{E} \quad \text{(in general)} \\
&= \frac{60}{120} \quad \text{(in this case)} \\
&= 0.5 \quad \text{amperes}
\end{aligned}
$$

So a circuit with a 3 amp fuse will safely take not more than six such bulbs at most.

Use the formula in its general form,

$$W = IE$$

combined with Ohm's law

$$\frac{E}{I} = R$$

and it tells us how to reduce transmission loses in power lines. Briefly, if P is the power delivered to the load at an emf E, the power lost on the way is

$$\left( \frac{P}{E} \right)^2 R_t$$

where $R_t$ is the resistance of the transmission lines.[1]

So if we double the working voltage and halve the current, we can deliver the same power with a quarter the transmission loss.

Although the letters are different, the equations

$$\frac{E}{I} = R \qquad \text{and} \qquad \frac{d}{t} = s$$

are both examples of the same mathematical model, $a/b = c$. Simply by changing the units, we have used this model for two very different jobs. The change of letters is simply to help us remember what, in this particular use, they represent.

Here is another use for the same model, this time for the refraction of light when passing from one transparent medium to another.

---

[1]Suppose that the power to be delivered to the load is P. Using $W = IE$ in this case gives

$$P = IE = I^2R.$$

The current I in the transmission line is the same as that in the load, so

$$I = \frac{P}{E} \quad .$$

So the power lost in the transmission lines is

$$I^2 R_t$$

$$= \left( \frac{P}{E} \right)^2 R_t$$

where $R_t$ is the resistance of the transmission lines.

$$\frac{\sin i}{\sin r} = \mu$$

where i is the angle of incidence, r the angle of refraction, and $\mu$ a constant called the *refractive index* for the two media. This model is a foundation for all the optical sciences, and an important contributor to all the other sciences that depend on microscopes, telescopes, cameras, and other optical instruments. We are now using the model a/b = c at two levels because both numerator and denominator are themselves ratios. These trigonometrical functions, sine and cosine, provide models that are used in many other ways, such as surveying, navigation, and electronic theory. I think that versatility of this kind is to be found only in mathematical models.

So, two of the ways in which mathematics now appears as a special case of the functioning of human intelligence are: (a) The use of mathematical models to make predictions, sometimes to achieve goal states, which would not be possible without them; and (b) Making explicit the multi-purpose nature of these models.

We cannot use a road map as a circuit diagram, or as anything else than a road map. Nor is a street map of Coventry any use to me when I visit Athens, Georgia. But as we have seen,

$$\frac{a}{b} = c$$

is a model that can be used for many different kinds of object and event.

## Uses of Mathematics in Group 2

As was seen in Chapter 8, co-operation between two or more persons requires that their various delta-one systems work together in some way, either to achieve a common goal, or compatible goals. And we noted in Chapter 8 that

> co-operation requires that the delta-one systems of the people concerned are using *complementary plans;* that is, plans which fit together to make an effective plan for the combination of their director systems working together. The success of any society depends on many and various people co-operating in many different ways for a great diversity of tasks, and this in turn depends on the availability of some way of arriving at complementary plans as and when required. The most successful way of doing this is by the existence of widely *shared schemas.* If individual plans are all derived from the same shared schema, this is a great step towards fitting them together. Co-operation also requires exchange of information, which is to say a *shared language* for the schema, its concepts and their relationships. (p. 113)

Mathematics provide excellent schemas for making co-operation possible. Where physical operands are involved, it is often only by the use of mathematics that we can give descriptions of physical objects and events that are exact enough for everyone's efforts to fit together. Manufacturers could not make nuts to fit bolts, pistons to fit cylinders, without the measurement functions of mathematics. It becomes more important when the co-operators are a distance apart, as in the case of tyre manufacturers in one place making tyres to fit wheels made somewhere else. And it is even more important when it is invisible properties of components that have to be fitted together. It would be no use for a designer of a television set to say to his supplier "I want a teeny weeny condenser, to fit in such-and-such a place in my circuit." He describes what he wants as an exact number of pico-farads.

The multi-purpose nature of mathematical schemas, discussed in group 1, comes in here also. By learning the same basic mathematics at school, people can use it when adult for co-operating in enterprises as various as designing and making motor cars, running railways, prescribing and giving the right doses for patients, making lenses for cameras, navigating ships that are out of sight of land and aircraft whose pilot cannot even see the ground, and putting satellites into orbit.

In all examples given so far, the operands are physical objects, so the uses for co-operation that I have described belong also to group 1: for achieving goals in the physical world. There is no imprecision here: these groups may and do intersect. But it is interesting to look also at uses in group 2 that do not involve group 1. That is, for co-operation based entirely on agreement about mathematical ideas, without the involvement of physical objects or events (except as carriers of information). One clear example of this is our monetary system, which makes possible extensive co-operation by the exchange of goods and services. When we buy a house, we do not give the seller any physical object of the same value. By writing a cheque, we transfer from our bank account to theirs what we may well describe as "a lot of money." But examined closely, it is just some rather large numbers. These numbers may be recorded on paper or in computers, but those are only symbols. It is the agreed meanings for these symbols that gives them their value, so that the receiver of the purchase price knows that he can acquire other useful objects or services by asking the bank to subtract certain numbers from his own bank account, and add the same numbers to the account of those who supply these goods or services.

If we borrow money for buying this house, this act of co-operation again depends on agreement about the payment of interest. The bank (or in the United Kingdom, very likely a building society) that lends us the money has in turn been lent this money by a large number of depositors. So large-scale co-operation is involved, with many individual transactions. These all fit together only because every single one is a particular case of the general relationship given by the formula

$$I = \frac{PTR}{100}$$

(I have simplified matters slightly by using simple interest. Compound interest is based on the same formula.) Here we have another example of the unifying power of mathematical ideas. And if you think that this is rather a simple example, may I suggest that you try a conceptual analysis of the ideas involved. Mathematical statements are so condensed that we tend to overlook the concentration of ideas ivolved.

Apart from money, other examples of the use of mathematics in ways that do not involve physical operands are harder to find than of those that do. But one such example is provided by elections, when by counting votes we agree who we shall be governed by. The mathematical model is very simple, but without it an important development in our social institutions would not be possible. Here, as elsewhere, a mathematical model is a necessary, but not a sufficient, condition for co-operation.

## Uses of Mathematical Schemas in Group 3

It was pointed out near the end of Chapter 8 that as schemas expand, they enable us to make use of inputs that previously could not have been assimilated. They sensitize us to observations that previously we would have passed by, ignored, or forgotten. So even in modes 1 and 2 of schema building, they act as agents of their own growth. The organic quality of schemas was also described, using as analogy the growth of a seed.

But certain kinds of schema seem able to grow even without input of this kind. A seed is very dependent on outside nourishment for its development into a plant. A daffodil bulb is less so, but it still needs water; and more importantly, it uses up its own substance, taken in and stored in earlier years. But certain kinds of schema can grow in quite a different way. Not only do these enlarge themselves without input: the more they grow the more they seem to ''want to'' go on growing. Have I used ''want to'' too loosely here? Perhaps, but nevertheless, there do seem to be times when the process of reflecting on and developing a schema almost seems to take over from conscious volition. One is reminded of times when, as may sometimes be seen, a dog appears to be taking its owner for a walk!

Do some kinds of schema have this quality more than others? Can we know in advance how to construct a schema in such a way that it will have these qualities?

I think that the answer is ''Yes'' to both questions, and that mathematical schemas offer good examples. One of the reasons for this is the abstractness and generality of their concepts, as already illustrated; and another is that mathematical schemas are connected mainly by conceptual links rather than by associative links.

One of my favorite examples is that of indices, already discussed in Chapter 4 as an example of reflective extrapolation (pages 40ff.). Another is provided by the successive generalisations of the number system, starting from the natural numbers, to the integers, the rational numbers, the real numbers, and imaginary numbers. It is interesting to notice that some of these are named retrospectively, in contrast with the next stage of generalisation.

## WHAT KIND OF THEORY IS MATHEMATICS?

Having made in Chapter 9 the distinction between type 1 and type 2 theories, the question naturally follows: What kind of theory is mathematics? It also looks like an important question: For how can we usefully think about teaching it if we do not know what kind of a subject it is that we are trying to teach?

For a start, we can see that any theory about *teaching* mathematics will be of a different category from mathematics itself. Theories about the learning and teaching of mathematics are clearly type 2 theories because they are theories about how we construct mathematical theories; but how about mathematics itself?

Mathematics is not one of the natural sciences, and the ultimate appeal is not to experiment. It was Einstein[2] who put the question "How can it be that mathematics, a product of human thought independent of experience, is so admirably adapted to the objects of reality?" Although in its very early stages it makes use of mode 1 building and testing (e.g., in the building of the concept of order and in the initial construction of the natural numbers) it rapidly abandons mode 1 and relies entirely on modes 2 and 3. Thus, correct or incorrect predictions of physical events play no part in confirming or infiming a mathematical theory, as they do a major part for other type 1 theories. But the discovery of an internal inconsistency would refute a mathematical theory. The discovery that new ideas were consistent with the accepted body of mathematical knowledge would help to confirm them; and a demonstration that they were a necessary consequence of certain parts of this accepted knowledge would constitute a proof, in the mathematical sense.

Although mathematics is not itself one of the natural sciences, it can be regarded as a conceptual kit of great generality and versatility, so valuable to anyone who wants to construct a scientific theory as to be almost indispensable. Did I say "almost"? Francis Bacon wrote: "For many parts of nature can neither be invented with sufficient subtlety, nor demonstrated with sufficient perspicuity nor accommodated into use with sufficient dexterity without the aid and intervention of mathematics." Likewise, Jeans: "All the pictures which science now

---

[2]This quotation, and those from Bacon, Jeans, and Galileo, are to be found as chapter headings in two books by Morris Kline, as follows. The Einstein quotation is at the beginning of Chapter 27; Bacon, Chapter 21; and Galileo, Chapter 13, in *Mathematics and the Physical World* (1960), London:Murray. The Jeans quote is at the beginning of Chapter 25 in *Mathematics: A Cultural Approach* (1962), Reading, MA: Addison-Wesley.

.

draws of nature and which alone seem capable of according with observational fact are mathematical pictures.'' A few examples have already been given. A collection that aimed at anything like comprehensiveness would look like a library of science.

So I suggest that we regard mathematics as a theory of a unique kind, having all the characteristics of a type 1 theory except mode 1 testing. It is the mental stuff of which type 1 theories are made; or to put this differently, it is pure form, of a kind that when given suitable content becomes a scientific theory. Give it a variety of contents, and we get a variety of scientific theories. I have said on many occasions that I regard mathematics as a particularly pure and concentrated example of the functioning of human intelligence. This suggests that it is a kind of essence, which when combined suitably with other ingredients communicates its qualities to the whole of the end product.

It is still hard to say why this should be. I still have not answered Einstein's question, so I offer the following as a beginning. Here is another quotation, this time from Galileo. ''Where our senses fail us, reason must step in.'' With the help of our senses, we perceive regularities of our physical environment. These regularities are embodied in what I have called *primary concepts*. Next, by the use of our intelligence, we find regularities among these regularities—we form secondary concepts. In mathematics, we repeat this process to form more ab-stract concepts, representing regularities of great generality, and relations be-tween these. All this time we are getting further away from what is accessible to our senses; yet, paradoxically, we seem to be getting closer to the essential nature of the universe. For when they are given content by using them to formulate scientific theories, these highly abstract mathematical concepts enable us to achieve results in our physical environment in ways that I myself continue to find astonishing whenever I think about them.

The foregoing, I hope, goes a little way toward explaining the relation of mathematics to the natural sciences; but we have not yet explained its uses in group 2, for co-operation. The uses for co-operation in actions on physical operands offer no great difficulty because the shared schemas from which the complementary plans arise are mathematically based in the way already de-scribed. But what about those concerned with non-physical operands, such as the systems of currency, buying and selling, borrowing and lending, and monetary exchange, which are so much a part of today's world? Money in particular is becoming more and more a matter of electronically transmitted and stored sym-bols, giving rise to paper print-outs or displays on cathode ray screens. Yet unless all these transactions are co-ordinated in ways about which everyone can agree, there would be chaos.

Once again, it is mathematics that comes to the rescue, providing us with systems for accountancy at all levels from personal to international; with ex-change rates between currencies that (at a given time) are agreed by all; and with ways of co-ordinating the transactions of those who want to borrow money, those who are willing to lend it, and those who act as intermediaries. And if we seek for the qualities that enable it to do so, these include its great internal consistency, which serves as a basis of wide agreement. The appeal is, above all, to reason. I do not suggest that reason is sufficient as a sole basis for co-operation— even more necessary is goodwill. But given the will to co-operate, human reason as embodied in mathematics provides, in many areas, a powerful help in finding the means.

# Relational Understanding and Instrumental Understanding[1]

## FAUX AMIS

*Faux amis* is a term used by the French to describe words which are the same, or very alike, in two languages, but whose meanings are different. For example:

| French word | Meaning in English |
|---|---|
| histoire | story, not history |
| libraire | bookshop, not library |
| chef | head of any organisation, not only chief cook |
| agrément | pleasure or amusement, not agreement |
| docteur | doctor (higher degree) not medical practitioner |
| médecin | medical practitioner, not medicine |
| parent | relations in general, including parents |

One gets *faux amis* between English as spoken in different parts of the world. An Englishman asking in America for a biscuit would be given what we call a scone. To get what we call a biscuit, he would have to ask for a cookie. And between English as used in mathematics and in everyday life there are such words as field, group, ring, ideal.

---

[1]Reprinted from "Mathematics Teaching," the Bulletin of the Association of Teachers of Mathematics No. 77 December, 1976.

A person who is unaware that the word he is using is a *faux ami* can make inconvenient mistakes. We expect history to be true, but not a story. We take books without paying from a library, but not from a bookshop; and so on. But in the foregoing examples there are cues which might put one on guard: difference of language, or of country, or of context.

If, however, the same word is used in the same language, country and context, with two meanings whose difference is non-trivial but as basic as the difference between the meaning of (say) 'histoire' and 'history,' which is a difference between fact and fiction, one may expect serious confusion.

Two such words can be identified in the context of mathematics; and it is the alternative meanings attached to these words, each by a large following, which in my belief are at the root of many of the difficulties in mathematics education today.

One of these is 'understanding'. It was brought to my attention some years ago by Stieg Mellin-Olsen, of Bergen University, that there are in current use two meanings of this word. These he distinguishes by calling them 'relational understanding' and 'instrumental understanding.' By the former is meant what I have always meant by understanding, and probably most readers of this article: knowing both what to do and why. Instrumental understanding I would until recently not have regarded as understanding at all. It is what I have in the past described as 'rules without reasons,' without realising that for many pupils *and their teachers* the possession of such a rule, and ability to use it, was what they meant by 'understanding.'

Suppose that a teacher reminds a class that the area of a rectangle is given by $A = L \times B$. A pupil who has been away says he does not understand, so the teacher gives him an explanation along these lines. "The formula tells you that to get the area of a rectangle, you multiply the length by the breadth." "Oh, I see," says the child, and get on with the exercise. If we were now to say to him (in effect) "You may think you understand, but you don't really," he would not agree. "Of course I do. Look; I've got all these answers right." Nor would he be pleased at our de-valuing of his achievement. And with his meaning of the word, he does understand.

We can all think of examples of this kind: 'borrowing' in subtraction, 'turn it upside down and multiply' for division by a fraction, 'take it over to the other side and change the sign,' are obvious ones; but once the concept has been formed, other examples of instrumental explanations can be identified in abundance in many widely used texts. Here are two from a text used by a former direct-grant grammar school, now independent, with a high academic standard.

*Multiplication of fractions* To multiply a fraction by a fraction, multiply the two numerators together to make the numerator of the product, and the two denominators to make its denominator.

$$E.g., \quad \frac{2}{3} \text{ of } \frac{4}{5} = \frac{2 \times 4}{3 \cdot 5} = \frac{8}{15}$$

$$\frac{3}{5} \text{ of } \frac{10}{13} = \frac{30}{65} = \frac{6}{13}$$

The multiplication sign $\times$ is generally used instead of the word 'of.'

*Circles* The circumference of a circle (that is its perimeter, or the length of its boundary) is found by measurement to be a little more than three times the length of its diameter. In any circle the circumference is approximately 3·1416 times the diameter which is roughly 3½ times the diameter. Neither of these figures is exact, as the exact number cannot be expressed either as a fraction or a decimal. The number is represented by the Greek letter $\pi$ (pi).

$$\text{Circumference} = \pi d \text{ or } 2\pi r$$
$$\text{Area} \qquad = \pi r^2$$

The reader is urged to try for himself this exercise of looking for and identifying examples of instrumental explanations, both in texts and in the classroom. This will have three benefits. (i) For persons like the writer, and most readers of this article, it may be hard to realise how widespread is the instrumental approach. (ii) It will help, by repeated examples, to consolidate the two contrasting concepts. (iii) It is a good preparation for trying to formulate the difference in general terms. Result (i) is necessary for what follows in the rest of the present section, while (ii) and (iii) will be useful for the others.

If it is accepted that these two categories are both well-filled, by those pupils and teachers whose goals are respectively relational and instrumental understanding (by the pupil), two questions arise. First, does this matter? And second, is one kind better than the other? For years I have taken for granted the answers to both these questions: briefly, 'Yes; relational.' But the existence of a large body of experienced teachers and of a large number of texts belonging to the opposite camp has forced me to think more about why I hold this view. In the process of changing the judgement from an intuitive to a reflective one, I think I have learnt something useful. The two questions are not entirely separate, but in the present section I shall concentrate as far as possible on the first: does it matter?

The problem here is that of a mis-match, which arises automatically in any *faux ami* situation, and does not depend on whether A or B's meaning is 'the right one.' Let us imagine, if we can, that school A send a team to play school B at a game called 'football,' but that neither team knows that there are two kinds (called 'association' and 'rugby'). School A plays soccer and has never heard of rugger, and vice versa for B. Each team will rapidly decide that the others are crazy, or a lot of foul players. Team A in particular will think that B uses a mis-shapen ball, and commit one foul after another. Unless the two sides stop and talk about what game they think they are playing at, long enough to gain some mutual understanding, the game will break up in disorder and the two teams will never want to meet again.

Though it may be hard to imagine such a situation arising on the football field, this is not a far-fetched analogy for what goes on in many mathematics lessons, even now. There is this important difference, that one side at least cannot refuse to play. The encounter is compulsory, on five days a week, for about 36 weeks a year, over 10 years or more of a child's life.

Leaving aside for the moment whether one kind is better than the other, there are two kinds of mathematical mis-matches which can occur.

1. Pupils whose goal is to understand instrumentally, taught by a teacher who want them to understand relationally.
2. The other way about.

The first of these will cause fewer problems *short-term* to the pupils, though it will be frustrating to the teacher. The pupils just 'won't want to know' all the careful ground-work he gives in preparation for whatever is to be learnt next, nor his careful explanations. All they want is some kind of rule for getting the answer. As soon as this is reached, they latch on to it and ignore the rest.

If the teacher asks a question that does not quite fit the rule, of course they will get it wrong. For the following example I have to thank Mr. Peter Burney, at that time a student at Coventry College of Education on teaching practice. While teaching area he became suspicious that the children did not really understand what they were doing. So he asked them: ''What is the area of a field 20 cms by 15 yards?'' The reply was: ''300 square centimetres.'' He asked: ''Why not 300 square yards?'' Answer: ''Because area is always in square centimetres.''

To prevent errors like the above the pupils need another rule (or, of course, relational understanding), that both dimensions must be in the same unit. This anticipates one of the arguments which I shall use against instrumental understanding, that it usually involves a multiplicity of rules rather than fewer principles of more general application.

There is of course always the chance that a few of the pupils will catch on to what the teacher is trying to do. If only for the sake of these, I think he should go on trying. By many, probably a majority, his attempts to convince them that being able to use the rule is not enough will not be well received. 'Well is the enemy of better,' and if pupils can get the right answers by the kind of thinking they are used to, they will not take kindly to suggestions that they should try for something beyond this.

The other mis-match, in which pupils are trying to understand relationally but the teaching makes this impossible, can be a more damaging one. An instance which stays in my memory is that of a neighbour's child, then seven years old. He was a very bright little boy, with an I.Q. of 140. At the age of five he could read *The Times,* but at seven he regularly cried over his mathematics homework.

His misfortune was that he was trying to understand relationally teaching which could not be understood in this way. My evidence for this belief is that when I taught him relationally myself, with the help of Unifix, he caught on quickly and with real pleasure.

A less obvious mis-match is that which may occur between teacher and text. Suppose that we have a teacher whose conception of understanding is instrumental, who for one reason or other is using a text which aim is relational understanding by the pupil. It will take more than this to change his teaching style. I was in a school which was using my own text (Skemp, 1962–69), and noticed (they were at Chapter 1 of Book 1) that some of the pupils were writing answers like 'the set of {flowers}.'

When I mentioned this to their teacher (he was head of mathematics) he asked the class to pay attention to him and said: "Some of you are not writing your answers properly. Look at the example in the book, at the beginning of the exercise, and be sure you write your answers exactly like that."

Much of what is being taught under the description of 'modern mathematics' is being taught and learnt just as instrumentally as were the syllabi which have been replaced. This is predictable from the difficulty of reconstructing our existing schemas.[2] To the extent that this is so, the innovations have probably done more harm than good, by introducing a mis-match between the teacher and the aims implicit in the new content. For the purpose of introducing ideas such as sets, mappings and variables is the help which, rightly used, they can give to relational understanding. If pupils are still being taught instrumentally, then a 'traditional' syllabus will probably benefit them more. They will at least acquire proficiency in a number of mathematical techniques which will be of use to them in other subjects, and whose lack has recently been the subject of complaints by teachers of science, employers and others.

Near the beginning I said that two *faux amis* could be identified in the context of mathematics. The second one is even more serious; it is the word 'mathematics' itself. For we are not talking about better and worse teaching of the same kind of mathematics. It is easy to think this, just as our imaginary soccer players who did not know that their opponents were playing a different game might think that the other side picked up the ball and ran with it because they could not kick properly, especially with such a mis-shapen ball. In which case they might kindly offer them a better ball and some lessons on dribbling.

It has taken me some time to realise that this is not the case. I used to think that maths teachers were all teaching the same subject, some doing it better than others. I now believe that *there are two effectively different subjects being taught under the same name, 'mathematics.'* If this is true, then this difference matters beyond any of the differences in syllabi which are so widely debated. So I would like to try to emphasise the point with the help of another analogy.

---

[2]See Chapter 3, this volume.

Imagine that two groups of children are taught music as a pencil-and-paper subject. They are all shown the five-line stave, with the curly 'treble' sign at the beginning; and taught that marks on the lines are called E, G, B, D, F. Marks between the lines are called F, A, C, E. They learn that a line with an open oval is called a minim, and is worth two with blacked-in ovals which are called crotchets, or four with blacked-in ovals and a tail which are called quavers, and so on—musical multiplication tables if you like. For one group of children, all their learning is of this kind and nothing beyond. If they have a music lesson a day, five days a week in school terms, and are told that it is important, these children could in time probably learn to write out the marks for simple melodies such as *God Save the Queen* and *Auld Lang Syne,* and to solve simple problems such as 'What time is this in?' and 'What key?', and even 'Transpose this melody from C major to A major.' They would find it boring, and the rules to be memorised would be so numerous that problems like 'Write a simple accompaniment for this melody' would be too difficult for most. They would give up the subject as soon as possible, and remember it with dislike.

The other group is taught to associate certain sounds with these marks on paper. For the first few years these are audible sounds, which they make themselves on simple instruments. After a time they can still imagine the sounds whenever they see or write the marks on paper. Associated with every sequence of marks is a melody, and with every vertical set a harmony. The keys C major and A major have an audible relationship, and a similar relationship can be found between certain other pairs of keys. And so on. Much less memory work is involved, and what has to be remembered is largely in the form of related wholes (such as melodies) which their minds easily retain. Exercises such as were mentioned earlier ('Write a simple accompaniment') would be within the ability of most. These children would also find their learning intrinsically pleasurable, and many would continue it voluntarily, even after O-level or C.S.E.

For the present purpose I have invented two non-existent kinds of 'music lesson,' both pencil-and-paper exercises (in the second case, after the first year or two). But the difference between these imaginary activities is no greater than that between two activities which actually go on under the name of mathematics. (We can make the analogy closer, if we imagine that the first group of children were initially taught sounds for the notes in a rather half-hearted way, but that the associations were too ill-formed and un-organised to last.)

The above analogy is, clearly, heavily biased in favour of relational mathematics. This reflects my own viewpoint. To call it a viewpoint, however, implies that I no longer regard it as a self-evident truth which requires no justification: which it can hardly be if many experienced teachers continue to teach instrumental mathematics. The next step is to try to argue the merits of both points of view as clearly and fairly as possible; and especially of the point of view opposite to one's own. This is why the next section is called *Devil's Advocate.* In one way this only describes that part which puts the case for instrumental understanding.

But it also justifies the other part, since an imaginary opponent who thinks differently from oneself is a good device for making clearer to oneself why one does think that way.

## DEVIL'S ADVOCATE

Given that so many teachers teach instrumental mathematics, might this be because it does have certain advantages? I have been able to think of three advantages (as distinct from situational reasons for teaching this way, which will be discussed later).

1.  Within its own context, *instrumental mathematics is usually easier to understand;* sometimes much easier. Some topics, such as multiplying two negative numbers together, or dividing by a fractional number, are difficult to understand relationally. 'Minus times minus equals plus' and 'to divide by a fraction you turn it upside down and multiply' are easily remembered rules. If what is wanted is a page of right answers, instrumental mathematics can provide this more quickly and easily.

2.  *So the rewards are more immediate, and more apparent.* It is nice to get a page of right answers, and we must not under-rate the importance of the feeling of success which pupils get from this. Recently I visited a school where some of the children describe themselves as 'thickos.' Their teachers use the term too. These children need success to restore their self-confidence, and it can be argued that they can achieve this more quickly and easily in instrumental mathematics than in relational.

3.  Just because less knowledge is involved, *one can often get the right answer more quickly* and reliably by instrumental thinking than relational. This difference is so marked that even relational mathematicians often use instrumental thinking. This is a point of much theoretical interest, which I hope to discuss more fully on a future occasion.

The above may well not do full justice to instrumental mathematics. I shall be glad to know of any further advantages which it may have.

There are four advantages (at least) in relational mathematics.

1.  *It is more adaptable to new tasks.* Recently I was trying to help a boy who had learnt to multiply two decimal fractions together by dropping the decimal point, multiplying as for whole numbers, and re-inserting the decimal point to give the same total number of digits after the decimal point as there were before. This is a handy method if you know why it works. Through no fault of his own, this child did not; and not unreasonably, applied it also to division of decimals. By this method $4 \cdot 8 \div 0 \cdot 6$ came to $0 \cdot 08$. The same pupil had also learnt that if you know two angles of a triangle, you can find the third by adding the two given angles together and subtracting from $180°$. He got ten questions right this way (his teacher believed in plenty of practice), and went on to use the same method for finding the exterior angles. So he got the next five answers wrong.

I do not think he was being stupid in either of these cases. He was simply extrapolating from what he already knew. But relational understanding, by knowing not only what method worked but why, would have enabled him to relate the method to the problem, and possibly to adapt the method to new problems. Instrumental understanding necessitates memorising which problems a method works for and which not, and also learning a different method for each new class of problems. So the first advantage of relational mathematics leads to:

2. *It is easier to remember.* There is a seeming paradox here, in that it is harder to learn. It is certainly easier for pupils to learn that 'area of a triangle = ½ base × height' than to learn why this is so. But they then have to learn separate rules for triangles, rectangles, parallelograms, trapeziums; whereas relational understanding consists partly in seeing all of these in relation to the area of a rectangle. It is still desirable to know the separate rules; one does not want to have to derive them afresh everytime. But knowing also how they are inter-related enables one to remember them as parts of a connected whole, which is easier. There is more to learn—the connections as well as the separate rule—but the result, once learnt, is more lasting. So there is less re-learning to do, and long-term time taken may well be less altogether.

Teaching for relational understanding may also involve more actual content. Earlier, an instrumental explanation was quoted leading to the statement 'Circumference = πd'. For relational understanding of this, the idea of a proportion would have to be taught first (among others), and this would make it a much longer job than simply teaching the rules as given. But proportionality has such a wide range of other applications that it is worth teaching on these grounds also. In relational mathematics this happens rather often. Ideas required for understanding a particular topic turn out to be basic for understanding many other topics too. Sets, mappings and equivalence are such ideas. Unfortunately, the benefits which might come from teaching them are often lost by teaching them as separate topics, rather than as fundamental concepts by which whole areas of mathematics can be inter-related.

3. *Relational knowledge can be effective as a goal in itself.* This is an empiric fact, based on evidence from controlled experiments using non-mathematical material. The need for external rewards and punishments is greatly reduced, making what is often called the 'motivational' side of a teacher's job much easier. This is related to:

4. *Relational schemas are organic in quality.* This is the best way I have been able to formulate a quality by which they seem to act as an agent of their own growth. The connection with 3 is that if people get satisfaction from relational understanding, they may not only try to understand relationally new material which is put before them, but also actively seek out new material and explore new areas, very much like a tree extending its roots or an animal exploring new territory in search of nourishment. To develop this idea beyond the level of an analogy is beyond the scope of the present paper, but it is too important to leave out.

If the above is anything like a fair presentation of the cases for the two sides, it would appear that while a case might exist for instrumental mathematics short-term and within a limited context, long-term and in the context of a child's whole education it does not. So why are so many children taught only instrumental mathematics throughout their school careers? Unless we can answer this, there is little hope of improving the situation.

An individual teacher might make a reasoned choice to teach for instrumental understanding on one or more of the following grounds.

1.   That relational understanding would take too long to achieve, and to be able to use a particular technique is all that these pupils are likely to need.

2.   That relational understanding of a particular topic is too difficult, but the pupils still need it for examination reasons.

3.   That a skill is needed for use in another subject (e.g., science) before it can be understood relationally with the schemas presently available to the pupils.

4.   That he is a junior teacher in a school where all the other mathematics teaching is instrumental.

All of these imply, as does the phrase 'make a reasoned choice,' that he is able to consider the alternative goals of instrumental and relational understanding on their merits and in relation to a particular situation. To make an informed choice of this kind implies awareness of the distinction, and relational understanding of the mathematics itself. So nothing else but relational understanding can ever be adequate for a teacher. One has to face the fact that this is absent in many who teach mathematics; perhaps even a majority.

Situational factors which contribute to the difficulty include:

1.   *The backwash effect of examinations.* In view of the importance of examinations for future employment, one can hardly blame pupils if success in these is one of their major aims. The way pupils work cannot but be influenced by the goal for which they are working, which is to answer correctly a sufficient number of questions.

2.   *Over-burdened syllabi.* Part of the trouble here is the high concentration of the information content of mathematics. A mathematical statement may condense into a single line as much as in another subject might take over one or two paragraphs. By mathematicians accustomed to handling such concentrated ideas, this is often overlooked (which may be why most mathematics lecturers go too fast). Non-mathematicians do not realise it at all. Whatever the reason, almost all syllabi would be much better if much reduced in amount so that there would be time to teach them better.

3.   *Difficulty of assessment* of whether a person understands relationally or instrumentally. From the marks he makes on paper, it is very hard to make valid inference about the mental processes by which a pupil has been led to make them; hence the difficulty of sound examining in mathematics. In a teaching situation, talking with the pupil is almost certainly the best way to find out; but in a class of over 30, it may be difficult to find the time.

4. *The great psychological difficulty for teachers of reconstructing their existing and longstanding schemas,* even for the minority who know they need to, want to do so, and have time for study.

From a recent article discussing the practical, intellectual and cultural value of a mathematics education (and I have no doubt that he means relational mathematics!) by Sir Hermann Bondi (1976), I take these three paragraphs. (In the original, they are not consecutive.)

> So far my glowing tribute to mathematics has left out a vital point: the rejection of mathematics by so many, a rejection that in not a few cases turns to abject fright.

> The negative attitude to mathematics, unhappily so common, even among otherwise highly-educated people, is surely the greatest measure of our failure and a real danger to our society.

> This is perhaps the clearest indication that something is wrong, and indeed very wrong, with the situation. It is not hard to blame education for at least a share of the responsibility; it is harder to pinpoint the blame, and even more difficult to suggest new remedies. (all on p. 8)

If for 'blame' we may substitute 'cause,' there can be small doubt that the widespread failure to teach relational mathematics—a failure to be found in primary, secondary and further education, and in 'modern' as well as 'traditional' courses—can be identified as a major cause. To suggest new remedies is indeed difficult, but it may be hoped that diagnosis is one good step towards a cure. Another step will be offered in the next section.

## A THEORETICAL FORMULATION[3]

There is nothing so powerful for directing one's actions in a complex situation, and for co-ordinating one's own efforts with those of others, as a good theory. All good teachers build up their own stores of empirical knowledge, and have abstracted from these some general principles on which they rely for guidance. But while their knowledge remains in this form it is largely still at the intuitive level within individuals, and cannot be communicated, both for this reason and because there is no shared conceptual structure (schema) in terms of which it can be formulated. Were this possible, individual efforts could be integrated into a unified body of knowledge which would be available for use by newcomers to the profession. At present most teachers have to learn from their own mistakes.

---

[3]This chapter was written before I had completed my new model of intelligence, described in Chapter 8. It represents an important step in my thinking—the relation between schemas and plans of action (see Chapter 9).

For some time my own comprehension of the difference between the two kinds of learning which lead respectively to relational and instrumental mathematics remained at the intuitive level, though I was personally convinced that the difference was one of great importance, and this view was shared by most of those with whom I discussed it. Awareness of the need for an explicit formulation was forced on me in the course of two parallel research projects; and insight came, quite suddenly, during a recent conference. Once seen it appears quite simple, and one wonders why I did not think of it before. But there are two kinds of simplicity: that of naivity; and that which, by penetrating beyond superficial differences, brings simplicity by unifying. It is the second kind which a good theory has to offer, and this is harder to achieve.

A concrete example is necessary to begin with. When I went to stay in a certain town for the first time, I quickly learnt several particular routes. I learnt to get between where I was staying and the office of the colleague with whom I was working; between where I was staying and the university refectory where I ate; between my friend's office and the refectory; and two or three others. In brief, I learnt a limited number of fixed plans by which I could get from particular starting locations to particular goal locations.

As soon as I had some free time, I began to explore the town. Now I was not wanting to get anywhere specific, but to learn my way around, and in the process to see what I might come upon that was of interest. At this stage my goal was a different one; to construct in my mind a cognitive map of the town.

These two activities are quite different. Nevertheless they are, to an outside observer, difficult to distinguish. Anyone seeing me walk from A to B would have great difficulty in knowing (without asking me) which of the two I was engaged in. But the most important thing about an activity is its goal. In one case my goal was to get to B, which is a physical location. In the other it was to enlarge or consolidate my mental map of the town, which is a state of knowledge.

A person with a set of fixed plans can find his way from a certain set of starting points to a certain set of goals. The characteristic of a plan is that it tells him what to do at each choice point: turn right out of the door, go straight on past the church, and so on. But if at any stage he makes a mistake, he will be lost; and he will stay lost if he is not able to retrace his steps and get back on the right path.

In contrast, a person with a mental map of the town has something from which he can produce, when needed, an almost infinite number of plans by which he can guide his steps from any starting point to any finishing point, provided only that both can be imagined on his mental map. And if he does take a wrong turn, he will still know where he is, and thereby be able to correct his mistake without getting lost; even perhaps to learn from it.

The analogy between the foregoing and the learning of mathematics is close. The kind of learning which leads to instrumental mathematics consists of the learning of an increasing number of fixed plans, by which pupils can find their way from particular starting points (the data) to required finished points (the answers to the questions). The plan tells them what to do at each choice point, as in the concrete example. And as in the concrete example, *what has to be done next is determined purely by the local situation.* (When you see the post office, turn left. When you have cleared brackets, collect like terms.) There is no awareness of the overall relationship between successive stages, and the final goal. And in both cases, the learner is dependent on outside guidance for learning each new 'way to get there'.

In contrast, learning relational mathematics consists of building up a conceptual structure (schema) from which its possessor can (in principle) produce an unlimited number of plans for getting from any starting point within his schema to any finishing point. (I say 'in principle' because of course some of these paths will be much harder to construct than others.)

This kind of learning is different in several ways from instrumental learning.
1.   The means become independent of particular ends to be reached thereby.
2.   Building up a schema within a given area of knowledge becomes an intrinsically satisfying goal in itself.
3.   The more complete a pupil's schema, the greater his feeling of confidence in his own ability to find new ways of 'getting there' without outside help.
4.   But a schema is never complete. As our schemas enlarge, so our awareness of possibilities is thereby enlarged. Thus the process often becomes self-continuing, and (by virtue of 3) self-rewarding.

Taking again for a moment the role of devil's advocate, it is fair to ask whether we are indeed talking about two subjects, relational mathematics and instrumental mathematics, or just two ways of thinking about the same subject matter. Using the concrete analogy, the two processes described might be regarded as two different ways of knowing about the same town; in which case the distinction made between relational and instrumental understanding would be valid, but not that between instrumental and relational mathematics.

But what constitutes mathematics is not the subject matter, but a particular kind of knowledge about it. The subject matter of relational and instrumental mathematics may be the same: cars travelling at uniform speeds between two towns, towers whose heights are to be found, bodies falling freely under gravity, etc. etc. But the two kinds of knowledge are so different that I think that there is a strong case for regarding them as different kinds of mathematics. If this distinction is accepted, then the word 'mathematics' is for many children indeed a false friend, as they find to their cost.

# Goals of Learning
# and Qualities
# of Understanding[1]

The present chapter is offered as a synthesis of ideas that have been discussed previously in journal articles (Backhouse, 1978; Buxton, 1978b; Byers & Herscovics, 1977; Skemp, 1976; Tall, 1978) and at a meeting of the British Society for the Psychology of Learning Mathematics. I am much indebted to all of these contributors for their stimulating interchange of ideas.

## A NEW MODEL OF INTELLIGENCE

The proposed synthesis has taken place around a new model of intelligence which I offer as a replacement for current ones based on 'I.Q.' All the latter can do is to put individuals approximately into rank order with respect to intelligence. It does not tell us what it is for, how it works, or how we can help learners to make the best use of whatever intelligence they have. In the new model I try to do all these; and, not surprisingly, mathematics appears as an important special case.

As a preliminary to the synthesis, we therefore need the briefest possible outline of this new model, which is, in essence, a two-level cybernetic one. What follows is no more than a thumbnail sketch—a full exposition takes over 300 printed pages (Skemp, 1979a).

Our starting point is the observation that much, possibly most, human behaviour is goal-directed; together with the conjecture that, cumulatively, success

[1]Reprinted from *Mathematics Teaching*, no. 88, Sept. 1979, pp. 44–49.

in achieving our goals is a major factor of survival. For goal-directed activity operating on the physical environment, we have a director system, delta-one, which receives information about the present state of the operand (what is being acted on), compares this with a goal state, and with the help of a *plan* which it constructs from its available *schemas*, takes the operand from its present state to its goal state and keeps it there. We may if we like call delta-one a sensori-motor system.

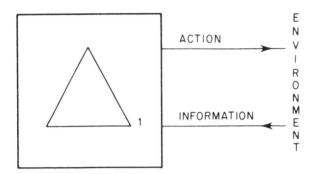

Delta-two is another director system, with a difference. Its operands are not in the outside environment, but in delta-one. They are not physical objects but mental objects. *The function of delta-two is to optimise the functioning of delta-one.* I prefer not to call delta-two a reflective system for reasons which will appear later. In a nutshell: *the job of delta-one is to direct physical actions, of many kinds. The job of delta-two is goal-directed mental activity,* also of many kinds, including learning, but not only learning. Learning includes the construction and testing by delta-two *within* delta-one of the schemas and plans which delta-one must have to do its job.

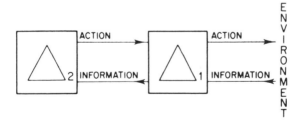

## HOW MANY KINDS OF UNDERSTANDING CAN WE USEFULLY DISTINGUISH?

With the help of this bare outline, we can make a start with trying to answer the above question, which emerges strongly from the present debate. My original single meaning was expanded to two at the suggestion of Mellin-Olsen (see Skemp, 1976), and to four by Byers and Herscovics (1977). Backhouse (1978) considers two sufficient. Buxton (1978b) describes four *levels* of understanding, which raises a new question. Are we talking about different points or regions alone a continuum, or about qualitative differences? He calls these levels 1. Rote, 2. Observational, 3. Insightful, 4. Formal. The first three he relates to the instrumental/relational distinction. Moreover, in the discussion at the B.S.Ps. L.M. meeting already mentioned, there was general agreement that Byers and Herscovics on the one hand, and Backhouse on the other, were using the term 'formal' with two importantly different meanings; one related to formal proof, and the other to 'form' in the sense, "This equation is in the form $y = mx + c$." Buxton's use of the term corresponds to the first of these two meanings; whereas for Backhouse, ". . . form is a way of representing a concept. . . ."

So we now have suggestions for seven different kinds and/or levels of understanding—with a tendency to proliferate, because of the creativity and individuality of those who are taking part in the discussion. But too many categories may be nearly as unhelpful as too few.

After considering all of these carefully, I think that we need to distinguish three kinds of understanding: Instrumental, Relational, Logical; and two modes of mental activity: Intuitive, Reflective. (Originally I called these "levels" of mental activity, but this was found sometimes to cause confusion with levels of abstraction.) "Logical" is simply a replacement for the first meaning of 'formal' so that we know which one we are talking about. I think that nearly all the rich assortment of ideas which we have been given can be fitted into this framework without doing them injustice. And we do not need to jettison the larger number of categories: they can be retained to give finer distinctions within the major categories, like 'sky blue' and 'ultramarine' within the category 'blue.'

I will start with the three different kinds of understanding. With one small change, I agree with Byers' and Herscovics' formulations. These are:

**Instrumental understanding** is the ability to apply an appropriate remembered rule to the solution of a problem without knowing why the rule works.

**Relational understanding** is the ability to deduce specific rules or procedures from more general mathematical relationships.

**Formal** [= Logical in my table] **understanding** is the ability to connect mathematical symbolism and notation with relevant mathematical ideas and to combine these ideas into chains of logical reasoning.

However, as Backhouse points out (my italics), ". . . we are unable to observe our pupils' feelings and schemas directly, and look for confirmatory behaviour. *As evidence that* A understands X, we accept the fact that A applies X in situations different (in greater or less degree) from that in which it was learned." So in Byers' and Herscovics' formulation, I would simply like to replace "is" in each case by ''is evidenced by.'' Note also that Backhouse's formulation applies equally well to instrumental and relational understanding. He writes, ''. . . situations different (in greater or less degree). . . .'' In situations which are not very different from each other, a rule will do. This corresponds to instrumental understanding. But if the situations are quite a lot different from each other, no simple rule will do, and amassing a collection of rules breaks down sooner or later. So it is better to be able, in the words of Byers and Herscovics, ''. . . to deduce specific rules or procedures from more general mathematical relationships.'' It is these relationships which are embodied in a relational schema.

What, if any, is the point of trying to make and formulate these distinctions? Are we simply engaged in an academic exercise in the pejorative sense of the word? The word ''usefully'' appears in the title of this section, and the onus is on me to justify it. I shall argue that these distinctions are not only useful, but essential to us as mathematics educators.

## SCHEMAS AND THE GOALS OF LEARNING

If it is accepted that ''To understand something means to assimilate it into an appropriate schema'' (Skemp, 1971) this implies that we can distinguish as many kinds of understanding as we can distinguish appropriate schemas. What decides whether a schema is appropriate or not? Here are two considerations (there may be others), the first fairly obvious, the second not always so.

(i) It has to be appropriate to the subject matter.
(ii) It has to be appropriate to the task in hand, that is, to the goal to be achieved.

As an example of the first, if my son asks me ''What is a differential?'' I need to know whether the subject matter is maths or motor cars before I can understand his meaning. But the same subject matter may be involved in different tasks, and these may imply important differences within what is superficially the same subject matter. A length of rope may seem a simple enough purchase, but there is more here than meets the eye. If it is for a halyard, I want rope which will not stretch, so that when the sail is hauled up tightly it stays that way; so I should buy, say, pre-stretched terylene. A climber, on the other hand, likes rope with good elasticity, so that in the event of a fall the impulsive tension is spread over a somewhat longer period; so he would be more likely to buy nylon. For a mooring I buy two kinds: a long length of non-floating rope from the buoy down to the holdfast, so that I do not create a danger to navigation; and a short length of floating rope for easy pick-up. For all these purposes, we want rope which does not rot. So if I ask a salesman for ''20 metres of 12 mm rope,'' he should ask what I want it for, or he may well sell me the wrong kind. He cannot do his job properly unless he knows the different kinds of uses for rope, and the qualities which are appropriate for these uses.

Likewise for mathematics, which we may indeed be trying to "sell" more actively than some of our pupils are wanting to "buy." Although the subject matter (mathematics) may be the same, for various persons and on various occasions the goals of learning may well be different, with the likelihood that different kinds of schema may be appropriate. So if we are trying to answer the question, "How many kinds of understanding can we usefully distinguish?" where "we" means ourselves as mathematical educators, then "usefully" now takes the meaning, "appropriate to the different kinds of learning goals set by our pupils and students." The distinction between various kinds of learning goal, which may appear superficially alike because involving the same subject matter, is one which we must have clearly available, or we cannot do our own jobs properly, any more than could a rope salesman to whom rope was just rope, without any thought that different persons might be acquiring it for different purposes. We must also remind ourselves that the goals of teachers and pupils may differ. Moreover, if we set up a learning situation which is conducive to one kind of learning goal, this will influence the kind of schemas pupils construct at least as much as, possibly more than, what is overtly being taught. So there will be a mis-match between the situationally-determined learning goals, and those which teachers may have in mind and/or verbally present.

We are now ready to think about three different kinds of learning goal, three associated kinds of schema, and thus three kinds of understanding.

## Instrumental Understanding

In schools, the *goal* of instrumental learning is to be able to give right answers, as many as possible, to questions asked by a teacher (verbally or on paper).

The *overt operands* (what are being manipulated) are symbols, mathematical and verbal, spoken or on paper.

The *hidden operand* is the teacher, in the first instance. The immediate goals of the pupil are to gain the teacher's approval and avoid the teacher's disapproval. Longer-term, the goal is approval by someone more remote—an examiner, a potential employer.

All of these operands (overt and hidden, short-term and long) are in the physical environment; so the activities described are delta-one activities. The teacher has asked a question and is waiting for an answer; an examiner has set a paper of questions, and the pupils have to answer as many as they can in a given time. The schemas formed by instrumental learning are short-term, the least and most quickly acquirable by which correct answers can be given: in other words *rules*, which we may regard as degenerate schemas. Pupils learn a set of these, each appropriate to a limited class of tasks. These rules are not entirely isolated—they can be combined, and applied in succession. But the mental structures acquired by instrumental learning have limited adaptability, because these rules are ways of manipulating symbols, and the connections are between symbols and not concepts. Application to substantially new situations requires conceptual connections, which is to say relational schemas.

## Relational Understanding

It is the construction of these relational schemas which are the *goals* of relational learning.

The *operands* may be newly encountered concepts, and the goal may be connecting these with an appropriate (relational) schema. Achievement of this goal is equivalent to relational understanding, and in the process the schema itself has undergone further development. Another kind of goal may be to deduce specific methods for particular problems, or specific rules for classes of tasks. Ability to do this is evidence of relational understanding. Yet another kind of goal is to improve the schemas we already have, by reflecting on them to make them more cohesive and better organised, and so more effective for the first and second kind of goal.

In all of these activities, the operands are concepts and schemas within delta-one, so *in relational learning, delta-two activity needs to be dominant*. And as Buxton has pointed out (1978a), this delta-two activity can be greatly interfered with if a teacher keeps asking questions! Discussion is quite another matter, but there must also be some quiet time for reflection, and people should not be required to speak if they do not want to. Alternation between these seems to be one of the best ways of encouraging delta-two activity. Discussion gives new material on which to reflect, requires one to formulate one's own ideas as clearly as possible, and shows up weaknesses in our thinking which by reflection we may remedy.

Symbols make an important contribution to all the activities described above. In reflection they act as combined handles and labels for their associated concepts, and we could not have any discussion without them. But the function of symbols is for manipulating and communicating mathematical concepts, and these are the true operands in relational mathematics.

Whereas the pleasure (if any) resulting from instrumental learning derives mainly from pleasing someone else—teacher, examiner—the delta-two activities involved in relational learning are a source of very personal pleasure. We ourselves know that the achievements of relational understanding can be very pleasurable; but many—perhaps a majority—of pupils unfortunately do not know this because they have never experienced it.

The goals of relational learning are long-term, in two ways.

(i) It takes longer to form mathematical concepts and construct relational schemas than it does to learn rules,

(ii) *but* long term it is the best way of pleasing teachers, passing examinations, satisfying employers, and achieving goals in a variety of future
situations for which mathematical methods are necessary.

Relational mathematics thus offers the best of both worlds. Unfortunately
many imposed learning situations are not conducive to this kind of learning.

## Logical Understanding[2]

It has been suggested that this category is unnecessary, being no different from
relational understanding. But (using Byers' and Herscovics' example) suppose
that a pupil writes something like

$$x + 3 = 7$$
$$= 7 - 3$$
$$= 4.$$

We ask why he writes this, and he answers in a way which convinces us that he
has relational understanding. (E.g. "The top line says that adding 3 to $x$ gives us
7, so to find what $x$ is we take the 3 away again.")

If we now point out to him the exact meaning of what he has written, and he
then re-writes it correctly, this would be evidence that he was capable of logical
understanding, which we had helped to bring about by causing him to reflect. By
". . . re-writes it correctly" we mean in an acceptable form which sets down a
valid sequence of logical inferences. If, however, he cannot see anything wrong
in what he has written, and continues to emphasise that he has given the right
answer (as evidenced by putting 4 for $x$ in the top line: "$4 + 3 = 7$, so I've got it
right, haven't I?") we might find it hard to convince him that we we're not just
being awkward. For there is indeed nothing wrong with his relational understanding, which relates to finding the value of $x$ which makes the equation a true
statement. What he lacks is logical understanding, which concerns the relationships of implication between the successive statements. We can thus discern
two kinds of not-having logical understanding: not-having it, and knowing that
one hasn't; and not-having it, and not knowing that one hasn't, because one has
not a schema of a kind which would enable one to recognise its presence or
absence.

Logical understanding is closely related to the difference between being convinced oneself, for which relational understanding is sufficient, and being able to
convince other people. To emphasise this I propose the following formulation.
**Logical understanding** is evidenced by the ability to demonstrate that what
has been stated follows *of logical necessity*, by a chain of inferences, from (i)

---

[2]I am indebted to Victor Byers and to members of the British Society for the Psychology of
Learning Mathematics at the seminar mentioned in the text, for useful discussion of this category.

the given premises, together with (ii) suitably chosen items from what is accepted as established mathematical knowledge (axioms and theorems). In this activity, the goals are not the acquisition of new concepts, nor the construction of new schemas, nor the devising of new methods to solve new problems. These must already be in existence as *operands* for the next stage, the *goal* now being to be sure that the schemas which have been constructed, the solutions which have been devised, are sound and accurate. The methods are what Bruner (1960) calls "the analytic apparatus of one's craft"—analysis, and the construction of chains of logical reasoning to produce what we call demonstration or proofs.

The primary operands are thus mathematical ideas which one already both has and understands. I think that we can also identify some secondary operands, namely one's fellow-mathematicians, whose critical evaluation we have to satisfy before whatever contributions we offer will be received into the body of accepted mathematical knowledge. So I would regard a mathematical proof or demonstration as corresponding to a replicable experiment in the natural sciences.

How does the term "formal" become attached to "proof" or "demonstration"? I suggest that a formal demonstration or proof is a method for exposing one's statement to the judgement of one's peers, by putting it in a form in which every implication is clearly shown, every theorem or axiom clearly referred to. "Form" here refers to the form of presentation, and "formal" means "conforming to accepted forms of presentation."

Most of us will also subject our ideas to self-criticism before making them public, formally or informally; and the construction of a proof which satisfies ourselves gives us confidence to do so. In some cases, all we want is to satisfy ourselves. But I would regard the satisfaction of criticisms and self-criticisms as secondary to the main goal, and indeed as aids to its achievement, the construction of ever more extensive and powerful mathematical knowledge, coherent, and without weaknesses or internal inconsistencies. This is clearly quite an advanced activity; yet as Buxton has pointed out (1978b), at one time,

> proofs of this sort, of a formal nature, were learnt by rote! Thus the most sophisticated level, valuable only to those who have achieved real insight and then recognised the need for proof is linked with the most primitive, thus firmly ensuing that the matter will not be understood!

One can hardly find a clearer example than this of the need to distinguish between different goals for learning, and the corresponding kinds of understanding which are required.

What kinds of schema are involved in this kind of understanding? Here I am working in my own frontier zone; perhaps others will help to take our knowledge further. It seems to me that the relationships involved are between the truth-values of statements. E.g. "*If* (it is true that) ABC is a triangle with AB, AC equal in length, *then* (it must also be true that) angles ABC and ACB are equal in size." "Then," in this context, means that it follows of *logical* necessity, not empirically, and not because teacher or any one else decrees it so. The reasoning involved is syllogistic: "In every triangle which has two sides equal in length, the angles opposite those sides are equal in size. ABC is a triangle which has two sides equal in length. Therefore . . ." So the schema involved seems to be one in which the concepts represent classes of statements, and the connections are those of logical implication. It would also seem that its location is in delta-two, since it is by means of this schema that delta-two operates on delta-one to produce schemas within delta-one which are free from inconsistencies. This might explain its inaccessibility to reflection, which is normally an activity in which delta-two becomes conscious of schemas within delta-one.

## THE INTUITIVE/REFLECTIVE DIMENSION

From the new model of intelligence we can also derive two modes of mental functioning. These do not relate to different kinds of understanding. On the contrary, they can occur in combination with all three of the kinds of understanding already discussed, as indicated below.

KINDS OF UNDERSTANDING

|  |  | INSTRUMENTAL | RELATIONAL | LOGICAL |
|---|---|---|---|---|
| MODES OF | INTUITIVE | $I_1$ | $R_1$ | $L_1$ |
| MENTAL ACTIVITY | REFLECTIVE | $I_2$ | $R_2$ | $L_2$ |

Though what follows may be recognised as a development of an earlier formulation, to save space I will not compare the two. In the present model, as was stated earlier, the job of delta-two is goal directed mental activity, including learning, but not only learning. "Goal-directed" does not imply "consciously goal-directed," either for delta-two or for delta-one. Many goal-directed activities of delta-one may take place unconsciously, such as those by which we maintain our state of balance while riding a bicycle and possibly holding a conversation at the same time; likewise for delta-two. Much learning takes place unconsciously; and since children learn from birth onwards, in the present model delta-two is assumed present from the beginning. (I conceive it as innate.) But we also know that reflective intelligence develops only gradually throughout childhood and adolescence. By this I mean the ability to make one's own concepts and mental processes objects of conscious attention, perhaps to describe them, perhaps also consciously and deliberately to change them. So what I now suggest is developing during this period is not delta-two itself, but the ability to shift our centre of consciousness from delta-one (the objects of consciousness being in the physical environment) to delta-two (the objects of consciousness being now in delta-one).

In the *intuitive mode* of mental activity, consciousness is centred in delta-one. In the *reflective mode*, consciousness is centred in delta-two. "Intuitive" thus refers to spontaneous processes, those within delta-one, in which delta-two takes part either not at all, or not consciously. "Reflective" refers to conscious activity by delta-two on delta-one.

With this formulation and with that of ealier sections, I can now try to justify the six categories in the $3 \times 2$ table above by offering examples of each.

## Category $I_1$

This corresponds to "rules withour reasons." Mechanical arithmetic comes into this category, but not the learning of skills to a degree where they can be performed fluently and with little or no thought, provided that when necessary they can be re-connected to an appropriate schema.

## Category $I_2$

This is an interesting one, since until I had set up this model I would have regarded reflective activity as necessarily relational. But consider this example.

Someone can differentiate the function

$$y = x^3$$

by using the rule

$$\begin{aligned} y &= x^n \\ \Rightarrow y' &= nx^{n-1}. \end{aligned}$$

He might have been told why some years ago, but he forgot it quite soon afterwards. Now he is asked to differentiate

$$y = \frac{1}{x^3}$$

His rule no longer applies; but on reflection, he remembers another rule:

$$\frac{1}{x^p} = x^{-p}.$$

By combining these rules, he gets

$$\begin{aligned} y &= \frac{1}{x^3} \\ \Leftrightarrow y &= x^{-3} \\ \Rightarrow y' &= -3x^{-4}. \end{aligned}$$

He has given the right answer without necessarily having relational understanding of either rule. What is more, logical understanding has been used for the inference, "If the first rule is correct, and if also the second rule is correct, then the result of using both rules in sequence will also be correct." Given that all he had to work with was an assortment of rules, he has tackled the problem quite intelligently. So we have an interesting implication, that reflective intelligence can be brought to bear on material in delta-one which is instrumental in quality. I now realise that this is what I have often been doing when, reflecting on mathematics I learnt many years ago, I came to realise that I had never understood it "properly" (i.e. relationally or logically).

## Category R₁

I think that this is what Bruner is referring to in the passages cited by Byers and Herscovics (1960):

> Intuition implies the act of grasping the meaning or significance or structure of a problem without explicit reliance on the analytic apparatus on one's craft. . . . It precedes proof; indeed, it is what the techniques of analysis and proof are designed to test and check.

In terms of the present model, intuitive understanding takes place when input to delta-one is directly assimilated to an appropriate schema, which thereby structures our perception of the problem. But intuition does not by itself imply understanding—the input may have activated inappropriate ideas, and our perception of the problem may on reflection prove to have been faulty. Hence the need for proof, as Bruner points out.

This emphasises the need to distinguish between intuition and insight. As Byers and Herscovics point out, "Many authors identify intuition with the Eureka phenomenon, the flash of sudden insight." But to do this is in my view mistaken. Insight (which I would equate with relational understanding) may indeed take place by an intuitive leap, but it may also be the result of prolonged reflection. And the result of an intuitive leap may, when tested by the techniques of analysis and proof, turn out not to have been a genuine insight.

## Category R₂

Using the calculus example again, let us imagine a student who not only knew the routines for differentiating the given function, and for rewriting $1/x^p$ without the fraction bar by using a negative index, but also had relational understanding of them. He might make an intelligent guess along the same lines as already described, and then proceed to verify this. To do this, he would need to relate these two routines to quite extensive mathematical schemas, including the binomial expansion for a negative exponent.

## Category $L_1$

Can logical understanding be identified at an intuitive as well as a reflective level? I believe so. Most children over ten, and many younger, would quickly say that "All dogs have tails. This animal is a dog; therefore, this animal has a tail" is a valid argument, whereas "All dogs have tails. This animal has a tail, therefore this animal is a dog" is not. Asked to say why the latter was false, they might give counter-examples, such as cats, squirrels. These might be taken as evidence of an intuitive awareness that a single counter-example was enough to reveal a false inference, but it might also be argued that all that was demonstrated was empirical knowledge. Stronger evidence for intuitive logic would be similar responses to statements like "All wonks have tails. . . ." I have not yet however been able to collect any mathematical examples for category $L_1$; contributions and suggestions will be welcomed.

## Category $L_2$

Understanding in this category would be evidenced by someone who responded to the example just given by drawing the appropriate Venn diagrams for the two cases. A more advanced pair of examples is provided by Russell's famous paradox. (In a certain village, the barber shaves everyone except those who shave themselves. Who shaves the barber?) $L_2$ understanding of this took a long time to achieve; yet we may assume that most people had an $L_1$ conviction that there was something wrong, from the start!

These apart, a major part of all mathematics at sixth form and university level comes into category $L_2$. Wherever an exercise or problem begins "Show that . . ." or "Prove that, . . ." $L_2$ understanding is needed and must be evidenced in the reply. And it is even more a key feature at research level, where "full mathematical rigour" means the same as a logically unassailable chain of implications.

Elsewhere (Skemp, 1971, Chapter 1 p. 15) I have distinguished between a logical and a psychological presentation, and have argued that it is a mistake to think that comprehension is best achieved by a strictly logical presentation. What is needed first is a presentation which is carefully planned psychologically, in order to bring about relational understanding. Until a learner understands what it is that one is trying to convince him about the truth of, then a logical argument is premature and out of place. But when relational understanding has been achieved, I think that there is a place for logical understanding even at quite an elementary level. Acceptance of a need both to give and to receive a logical proof says, by implication, "I don't expect you to believe this on my say-so. I hope to convince you that what I am saying follows of logical necessity from things we both agree about. If you will follow my chain of inferences step by step, I think that you will be convinced by the exercise of your own reason.'' This further implies that the final authority for accepting or not accepting a statement lies in the mathematics itself; which is something which, in ways appropriate to the level of the learner, I think we should be teaching from early years onwards.

# Communicating Mathematics: Surface Structures and Deep Structures[1]

The power of mathematics in enabling us to understand, predict, and sometimes to control events in the physical world lies in its conceptual structures—in everyday language, its organised networks of ideas. These ideas are purely mental objects: invisible, inaudible, and not easily accessible even to their possessor. Before we can communicate them, ideas must become attached to symbols. These have a dual status. Symbols are mental objects, about which and with which we can think. But they can also be physical objects—marks on paper, sounds—which can be seen or heard. These serve both as labels and as handles for communicating the concepts with which they are associated. Symbols are an interface between the inner world of our thoughts, and the outer, physical world.

These symbols do not exist in isolation from each other. They have an organisation of their own, by virtue of which they become more than a set of separate symbols. They form a symbol system. A symbol system consists of

a set of symbols    corresponding to          a set of concepts
together with
a set of relations    corresponding to         a set of relations
between the symbols                            between the concepts.

*What* we are trying to communicate are the conceptual structures. *How* we communicate these, or try to, is by writing or speaking symbols. The first are

[1]Reprinted from *Visible Language*, vol. XVI, no. 3, pp. 281–288.

what is most important. These form the *deep structures* of mathematics. But only the second can be transmitted and received. These form the *surface structures*. Even within our minds the surface structures are much more accessible, as the term implies. And to other people they are the only ones which are accessible at all. But the surface structures and the deep structures do not necessarily correspond, and this causes problems.

Here are some examples to illustrate the differences between a surface structure and a deep structure.

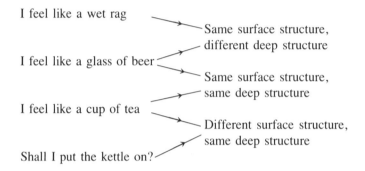

What has this to do with mathematics? At a surface level wet rags and cups of tea would seem to have little connection with mathematics. But at a deeper level, this distinction between surface structures and deep structures, and the relations between them, is of great importance when we start to think about the problems of *communicating* mathematics.

For convenience let us shorten these terms to $S$ for surface structure, $D$ for deep structure. $S$ is the level at which we write, talk, and even do some of our thinking. The trouble is that the structure of $S$ may or may not correspond well with the structure of $D$. And to the extent that it does not, $S$ is inhibiting $D$ as well as supporting it.

Let us look at some mathematical examples. We remember that a symbol system consists of:

(i)  a set of symbols, e.g.,  1   2   3   . . .

$\frac{1}{2}$  $\frac{3}{4}$      . . .

a   b   c   . . .

(ii) one or more relations on those symbols, e.g. order on paper (left/right, below/above); order in time, as spoken.

But since the essential nature of a symbol is that it represents something else—in this case a mathematical concept—we must add

(iii) such that these relations between the symbols represent, in some way, relations between the concepts.

So we must now examine what ways these are, in mathematics. Here is a simple example. (Note that 'numeral' refers to a symbol, 'number' refers to a mathematical concept.)

| *Symbols* | *Concepts* |
|---|---|
| (i)  1 2 3 . . . (numerals in this order) | the natural numbers |
| *Relations between symbols* | *Relations between concepts* |
| (ii)  is to the left of (on paper) before in time (spoken) | is less than |

This is a very good correspondence. It is of a kind which mathematicians call an isomorphism. Place value provides another well known example of a symbol system.

| *Symbols* | *Concepts* |
|---|---|
| (i)  1 2 3 . . . (numerals) | natural numbers |
| *Relations between symbols* | *Relations between concepts* |
| (ii)  numeral$_1$ is one place left of numeral$_2$. | number$_1$ is ten times number$_2$. |

By itself this is also a very clear correspondence. But taken with the earlier example, we find that we now have the same relationship between symbols, *is immediately to the left of,* symbolising two different relations between the corresponding concepts: *is one less than* and *is ten times greater than.* We might take care of this at the cost of changing the symbols, or introducing new ones; e.g., commas between numerals in the first example. But what about these?

23      2½      2a

These can all occur in the same mathematical utterance. And this is not just carelessness in choice of symbol systems; it is inescapable, because the available relations on paper or in speech are quite few: left/right, up/down, two dimensional arrays (e.g., matrices); big and small (e.g., $R$, $r$) What we can devise for the surface structure of our symbol system is inevitably much more limited than the enormous number and variety of relations between the mathematical concepts, which we are trying to represent by the symbol system.

Looking more closely at place value, we find in it further subtleties. Consider the symbol: 572. At the S level we have three numerals in a simple order relationship. But at the D level it represents

(i)  three numbers,           5     7     2
     corresponding to         $\updownarrow$     $\updownarrow$     $\updownarrow$
(ii) three powers of ten:    $10^2$   $10^1$   $10^0$

These correspond to the three locations of the numerals, in order from right to left.

(iii) three operations of multiplication: the number 5 multiplied by the number $10^2$ (= 100), the number 7 multiplied by the number $10^1$ (= 10), the number 2 multiplied by the number $10^0$ (= 1).

(iv) addition of these three products (5 hundreds, seven tens, two).

Of these four at D level, only the first is explicitly represented at S level by the numeral 572. The second is implied by the spatial relationships, not by any visible mark on the paper. And the third and fourth have no symbolic counterpart at all: they have to be deduced from the fact that the numeral has more than one digit.

Once one begins this kind of analysis, it becomes evident there is a huge and almost unexplored field—enough for several doctoral theses. For our present purposes, it is enough if we can agree that the surface structure (of the symbol system) and the deep structure (of the mathematical concepts) can at best correspond reasonably well, in limited areas, and for the most part correspond rather badly.

To help our thinking further in this difficult area, I would like to introduce two further ideas. The first comes from my new model of intelligence (1979a) and does not require any other parts of the theory. It is based on the well-known phenomenon of resonance. "The starting point is to suppose that conceptualised memories are stored within tuned structures, which, when caused to vibrate, give rise to complex wave patterns. . . . Sensory input which matches one of these wave patterns resonates with the corresponding tuned structure, or possibly several structures together, and thereby sets up the particular wave pattern of a certain concept." (page 134)

It is convenient at this stage to introduce the term *schema*, which is simply a shorter way of referring to a conceptual structure. A schema (i.e., a conceptual structure stored in memory) thus corresponds in this model to a particular tuned structure. We all have many of these tuned structures corresponding to our many available schemas, and sensory input is interpreted in terms of whichever one of these resonates with what is coming in. What is more, different structures may be thus activated by the same input in different people, and at different times in the same person. Different interpretations will then result. For example, the word 'field' will have quite different meanings according as it evokes resonances corresponding to the schemas in advanced mathematics, electromagnetism, cricket, agriculture, or general scholarship.

The second idea is due to Tall (1977) who has suggested that a schema can act as an attractor for incoming information. He took the idea from the mathematical theory of dynamic systems; but if we now combine it with the resonance model, we can offer an explanaion of how this attraction might take place. Sensory input will be structured, interpreted, and understood in terms of which ever resonant structure it activates. In some case, more than one resonant structure may be activated simultaneously, and we can turn our attention at will to one or the other. In others, one schema captures all the input. (This 'capture effect' is well known to radio engineers, who have put it to good use.)

So we may now synthesise the following ideas.

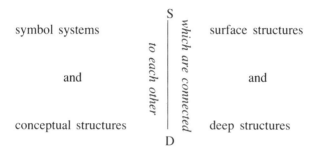

Note that in the above diagram each point represents not a single concept but a schema, in the same way as a dot on an airline map can represent a whole city— London, Atlanta, Rome.

How can this theoretical model help our thinking, and what are the practical consequences? All communication, written or oral, is necessarily into the symbol system at S. *To be understood mathematically, it must be attracted to D.* This requires that D is a stronger attractor than S. If it is not, *S will capture the input,* or most of it.

One of the advantages of a good model is that it points up some questions we should ask next. The first is clearly: What are the conditions for D to be a strong attractor? Another is: can D capture the input instead of S? If so what happens?

I will take the second first, briefly. If this were to happen, I think it would mean that all the mathematical activity was confined to a deep conceptual level, and was not 'escaping' to a symbolic level at all. This may not happen completely, but some of the high-powered mathematicians who taught me at university suggest only very limited escape to S!

Returning to the first question: what are the conditions for D to be a strong attractor? S has a built in advantage: all communicated input has to go there first. And for D there is a point of no return. In the years' long learning process, if the deep conceptual structures are not formed early on, they can never develop as attractors. For too many children, D is effectively not there. And if the D structure is absent or weak, all input will be assimilated to S: the effort to find some kind of structure is strong. So S will build up at the expense of D. But this guarantees problems, in view of the lack of internal consistency of S. This reveals a built-in advantage of D, that it *is* internally consistent. Of all subjects, mathematics is one of the most internally consistent and coherent. So if it gets well established, input to S will evoke more extensive and meaningful resonances in D than in S, and D will attract much of the input.

Doing mathematics involves the manipulation of certain mental objects, namely mathematical concepts, using symbols as combined handles and labels. But for many children (and adults) these objects are not there. So they learn to manipulate substitute objects: empty symbols, handles without anything attached, labels without contents. This in the long run is much more difficult to do, though unfortunately in the short run it may be easier to learn. The manipulation of mathematical concepts is helped by the nature of the concepts and schemas themselves, which give a feeling of intrinsic rightness or wrongness. This arises partly from the concepts themselves, whose individual properties contribute to how we use them and fit them together. More strongly, it comes from the schemas, which determine what are permissible and non-permissible mental actions within a given mathematical context.

The problems which so many have with mathematical symbols thus arise partly from the laconic, condensed, and often implicit nature of the symbols themselves; but largely also from the absence or weakness of the deep mathematical schemas which give the symbols their meaning. Like a referred pain, the location of the trouble is not where it is experienced. The remedy likewise lies mainly elsewhere, namely in the building up of the conceptual structures.

How can we help learners to do this? This is too large a question for a single paper, but here are some suggestions as starting points.

(i) Particularly in their early years we can give children as many physical embodiments as possible of the mathematical concepts which we want to help them to construct. As examples of units, tens, and hundreds, we can use single milk straws, bundles of ten of these, and bundles of ten bundles of ten. These correspond much more closely to the relevant mathematical concepts than do the associated symbols, and so the visual input will be attracted more strongly to the relevant parts of D than to S. In such cases, moreover, the input goes first to D, then to S, since the children are first presented with the physical embodiments of the concepts, and thereafter are asked to connect these with appropriate symbols.

(ii) By careful analysis of the mathematical structure to be acquired, we can sequence the presentation of new material in such a way that it can always be assimilated to a conceptual structure, and not just memorised in terms of symbolic manipulations. Many existing texts show no evidence that this has been done. (See Skemp, 1971, Chapter 2.)

(iii) Again in these important early years, it helps children if we stay longer with spoken language. The connection between thought and spoken words are initially much stronger than those between thoughts and written words or symbols. Spoken words are also much quicker and easier to produce. So in the early years of learning mathematics, we may need to resist pressures for children to have 'something to show' in the form of pages of written work.

(iv) It is often helpful to use informal, transitional notations as bridges to the formal, highly condensed notations of traditional mathematics. By allowing children to express their thoughts in their own ways to begin with, we are using symbols which are already well attached to their associated concepts. These ways of expression may often be lengthy, unclear, and differ between individuals. By experience of these disadvantages, and by discussion, children may gradually be led to the use of established mathematical symbolism in such a way that they experience its convenience and power for communicating and manipulating mathematical ideas.

# 15

# Symbolic Understanding[1]

In this chapter, I offer a further contribution to a series of discussions about the nature and varieties of mathematical understanding which has taken place in the past (Backhouse, 1978; Buxton, 1978b; Byers & Herscovics, 1977; Skemp, 1976; Tall, 1978). By 1978 seven categories had been proposed, which I subsequently suggested (see Chapter 13) could be re-arranged into a table showing three kinds of understanding and two modes of mental activity. However, as I was aware at the time, my analysis of formal understanding was incomplete, since the words 'form' and 'formal' are used with two distinct meanings, of which I only dealt with the first. (i) There is 'form' as in 'formal proof.' This is the meaning used by Buxton (1978) and I have already suggested (see Chapter 13) that we distinguish this one by calling it 'logical understanding.' (ii) There is 'form' as used in statements such as, "This equation can be written in the form $y = mx + c$." This is the meaning used by Backhouse (1978), and is also that in the first part of the definition given by Byers and Herscovics (1977): "Formal understanding is the ability to connect mathematical symbolism and notation with relevant mathematical ideas . . ." I now suggest that we distinguish this meaning by calling it *symbolic understanding*.

Recently, I have been emphasising that the achievement of new understanding gives new abilities (Skemp, 1980). So what can we do when we have symbolic understanding that we could not do before? The power of mathematical symbolism is a special case of the power of language, so we would expect these powers to be great. Here are ten (there may be others) which I listed some years

---

[1]Reprinted from *Mathematics Teaching* No. 99, June 1982, pp. 59–61.

ago (see Chapter 5). At that time I had not yet seen these in relation to the acquisition of a new kind of understanding, since none of the present series of discussions had taken place.

1. Communication.
2. Recording knowledge.
3. The formation of new concepts.
4. Making multiple classification straightforward.
5. Explanation.
6. Making possible reflective activity.
7. Helping to show structure.
8. Making routine manipulations automatic.
9. Recovering information and understanding.
10. Creative mental activity.

The powers conferred by symbolic understanding are immense, though we are so used to them that we tend to take them for granted. The task of acquiring it is also a considerable one, and we easily overlook the achievement of children in learning to speak their mother tongue with considerable mastery by the age of five. But we cannot overlook the difficulties which many children have in learning to understand mathematical symbolism.

There is a new factor to be taken account of here. In the earlier discussions, we were concerned with the assimilation of concepts to schemas. But in the case of symbolic understanding, 'symbolic' refers to a symbol *system,* not to a single symbol. A symbol system is a set of symbols corresponding to a set of concepts, together with relations between the symbols corresponding to relations between the concepts. For example, the symbols '2' and '3' taken separately refer to particular number-concepts. When we write them like this '$2^3$' we use two relationships between the symbols, one of size and one of position. At the conceptual level, these together specify an operand (the number 2) and an operation (raising to the power 3). If both symbols are the same size and on the same level, as in 23, the meaning is quite different. So we are now dealing with two schemas: the symbol system, and the structure of mathematical concepts. This suggests the provisional formulation: symbolic understanding is a mutual assimilation between a symbol system and an appropriate conceptual structure.

So now we are concerned not with the assimilation of concepts to schemas, of small entities to large ones, but with the mutual assimilation of two schemas, of two entities which are comparable in size and each of which has a structure of its own. A comparable event in the history of mathematics can be found in the great achievement of Descartes, who assimilated two major structures, geometry and algebra, to each other. When something like this happens, as well as an increase of power, there is also the possibility that one organisation, in this case a mental one, dominates the other. Whether or not this is desirable will vary in different instances. Since Descartes there has been, it appears to me, a progressive take-over in this partnership by the algebra. We can now find points *defined* as ordered pairs, triplets, or *n*-tuples of numbers; and books, whose titles say that they are about geometry, in which one finds nothing but algebra, without a single diagram or geometrical figure.

We may or may not think that this is good; and although I think myself that it is not, I accept the opposite as a tenable position. But I trust that none of us is happy with a relationship in which the conceptual structure is dominated by the symbol system, and mathematics is little or nothing more than the manipulating of symbols. The power of mathematics is in the ideas. In the right partnership symbols help us to use this power by helping us to make fuller use of these ideas. Yet the former situation is the way it is for all too many children.

Where there is isomorphism between the two structures, it may matter little which one dominates, either in an individual or collectively. Part of the success of algebraic geometry lies in the closeness of this isomorphism, so that each structure helps to increase our understanding of the other. But between the symbol systems and the conceptual structures of mathematics, we find local isomorphisms only. Overall there are many inconsistencies. For example, the spatial relationship, *is next on the left to,* means three different things in these three cases:

$$23 \qquad\qquad 2\frac{1}{2} \qquad\qquad 2a$$

Another example is the ordered pair of numerals (2,3), which can signify a rational number, a point in a plane, or a free vector. With the first meaning, we add like this: $(2,3) + (4,5) = (2 \times 5 + 3 \times 4, 3 \times 5)$. With the second meaning, we cannot add at all. With the third meaning, we add like this: $(2,3) + (4,5) = (2 + 4, 3 + 5)$ (which is the way some children find it more natural to add rational numbers). And this is not just carelessness in our choice of symbol systems. It is inescapable, because the available relations between symbols are few. Left and right, up and down, big and small (as in indices and suffixes), bold face and light (which we cannot speak, only print)—we soon run out of these. But the relations between mathematical concepts which we are trying to represent are many, and continue to increase as our knowledge advances.

So how can we help children to build up an increasing variety of meanings for the same symbols? How can we prevent them from becoming progressively more insecure in their ability to cope with the increasing number, complexity, and abstractness of the mathematical relations they are expected to learn?

As a help towards answering this question, I would like to make use of a model based on the well-known phenomenon of resonance (Skemp, 1979a).

The starting point is to suppose that conceptualised memories are stored within tuned structures which, when caused to vibrate, give rise to complex wave patterns. . . . Sensory input which matches one of these wave patterns resonates with the corresponding tuned structure, or possibly several structures together, and thereby sets up the particular wave pattern of a certain concept.

A schema, being a conceptual structure stored in memory, thus corresponds (in this model) to a particular, complex, tuned structure. We all have many of these, and sensory input will be interpreted in terms of whichever one of these resonates with what is coming in. What is more, for different people, different structures may be activated in this way—caused to vibrate—by the same input. Even for the same person and the same input, different structures may be activated at different times. Thus 'field' will cause different vibrations according as the schemas with which it resonates are mathematical, electro-magnetic, farming, academic, or cricket. Whichever schema resonates most easily *will attract the input.*

In the case we are discussing, there are two contenders: the symbol system and the conceptual structure.

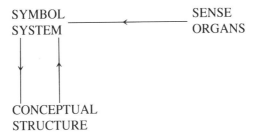

Since communication is by the utterance of symbols, all communication, whether verbal or written, first goes into a symbol system. To be understood relationally, it must be attracted to an appropriate conceptual structure. What is more, the input must be interpreted in terms of the relationships within the conceptual structure, rather than those of the symbol system. (E.g., 572 must be interpreted, not as a succession of three single-digit numbers, but as a single number formed by the sum $5 \times 10^2 + 7 \times 10 + 2$.) This requires: (i) that the conceptual structure is a stronger attractor than the symbol system; (ii) that the connections between the symbol system and the conceptual structure are strong enough for the input to go easily from the first to the second.

How can we help this to happen? I have five suggestions to offer—briefly, for each could usefully be expanded into a chapter of a book.

(1)   We have noted that the symbol system has a built-in advantage, that all communications necessarily go there first. And, for the conceptual structure, there is a point of no return. In the years long process of learning mathematics, if those conceptual structures are not formed early on, they will never get the chance to develop as attractors. The effort to find some kind of regularity is strong. If the conceptual structure is absent or weak, the input will be assimilated

to the symbol system. But this guarantees problems, for we have seen that the symbol system is inconsistent. Learning at this level may be easy short-term, but it becomes impossibly difficult long-term. In contrast, the conceptual structure *is* (or should be) internally consistent. Of all subjects, relational mathematics is one of the most internally consistent and coherent, so long-term it is much easier to learn and retain. Part of the answer is that by careful analysis of the mathematical concepts, we must sequence material in such a way that new material is presented which can always be assimilated *conceptually* (see Chapters 2 & 3; also Concept Maps section Chapter 9).

(2)   Especially in the early years, we can begin with physical embodiments of mathematical concepts and activities, so that the sensory input goes *first* to the conceptual structure and is *then* connected with its symbolic representation.

(3)   Again especially in these all-important early years, I think that we should stay with spoken language much longer. Recently I came upon a nice quotation from Sartre (1964): "On parle dans sa propre langue, on écrit en langue étrangère." The connections between thoughts and spoken words are initially much stronger than those between thoughts and written words, or thoughts and mathematical symbols. Spoken words are also much quicker and easier to produce. So in the early years we need to resist pressures to have 'something to show' in the form of pages of written work.

(4)   Some notations, such as the use of parentheses to denote the order of operations, can be seen to arise out of the needs of a situation (e.g., Kieran, 1979).

(5)   We should use transitional, informal notations as bridges to the formal, highly condensed notations of established mathematics. By allowing children to express thoughts in their own ways to begin with, we are using symbols already well-attached to their conceptual structure. These ways will probably be lengthy, ambiguous and different between individuals. By experiences of these disadvantages, and by discussion, children may be led gradually to the use of conventional notation in such a way that they experience its convenience and power.

In the light of the foregoing discussion, I offer the following revised formulation.

Symbolic understanding is a mutual assimilation between a symbol system and a conceptual structure, *dominated by the conceptual structure.*

Symbols are magnificent servants, but bad masters, because by themselves they do not understand what they are doing.

# Emotions and Survival
# in the Classroom

Mainstream psychology has tended either to ignore emotions, or to regard them as irrational influences, distractors, which disturb our normal thinking processes. This is in accordance with everyday usage: my Concise Oxford Dictionary defines *emotion* as 'agitation of mind, feeling; excited mental state.' However, I believe that the separation of cognitive from affective processes is an artificial one, which does not accurately reflect human experience. In particular, many students have reported that strong emotions have been aroused by their classroom experiences, and that these have strongly influenced their learning for better or worse. In Chapter 7, I made a first approach to the subject. In this chapter, I suggest how the model of intelligence outlined in Chapter 8 can be extended to include this important influence on our behaviour and learning.

## DO EMOTIONS HAVE A USEFUL FUNCTION?

Given that emotions are, subjectively, an important feature of human experience, it seems reasonable to ask whether we should indeed regard them as a disturbance of our normal thinking processes, in which case we should try to minimise their influence; or whether emotions have a useful function, in which case we need to know what this is.

I believe that this, like many other questions relating to human nature and activity, can only be understood in the context of evolution. This is another reason why I prefer a biological model to one based on analogies with the functioning of computers. So I shall begin with a wide-angle view of intelligence in relation to adaptation and survival.[1]

Any species that exists on earth today is here because over the centuries it has evolved physical, behavioural, and mental characteristics that are pro-survival. We have become dominant on this planet mainly because of one particular characteristic, our intelligence. Why is intelligence pro-survival? Because it gives us the ability to achieve our goal states in a variety of ways to suit a variety of circumstances. Intelligence shows not in behaviour itself, but in adaptive changes of behaviour. It has already been shown, in Chapter 11, that mathematics is an important source of these.

If the survival value of intelligence is adaptability, and the survival value of adaptability is that it enables us to achieve our goals, why is this last pro-survival? Because many of our goals are directly related to survival. In some cases, the connection is direct and obvious: eating, drinking, keeping the right body temperature. In others, perhaps, is is less direct. The survival value of knowledge does not show at once, and at the time we acquire it we may not even know what we shall use it for.

But, like money in the bank, it is good for a wide variety of purposes; and when our immediate needs are taken care of, the pursuit of knowledge has long-term survival value. In this, we are providing ourselves with a resource from which we can construct a variety of plans of action to serve needs as they arise. Some of these may not have been foreseen when the knowledge resources was being acquired. When at school I learned De Moivre's theorem (one of my favorites), little did I know that I would find it useful as an army signals officer. Although this elegant theorem is not usually regarded as having survival value, in this case the connection does not need spelling out.

The previous, necessarily brief, wide-angle view suggests that if we are looking for a possible answer to the question ''Do emotions serve any useful purpose?,'' a good place to begin might be by looking for some survival value in emotions.

At any time, our senses tell us of many changes of state in our surroundings. Some of these changes take us nearer to, or further from, our goal states. Collectively and statistically, these changes relate to our survival, whereas others are neutral. So it would be pro-survival in itself to have signals that call our attention to changes that do relate to goal states; and it would also be pro-survival if these signals were qualitatively different from other data reaching our consciousness because this would enhance their attention-getting quality. This may be recognised as a fair description of emotions. They are hard to ignore, and this is because they are calling our attention to matters that relate to our survival.

---

[1]The relation between intelligent learning and survival is discussed at greater length in Skemp, 1979a, Chapter 1.

## Emotions in Relation to Goals and Anti-Goals

The categories of emotion that follow are broad, and within each there are differentiations that I do not make here.

**Pleasure.**   Emotions in this category signal changes toward a goal state. We feel pleasure while eating, taking exercise, resting when tired, enjoying the company of friends. The first three of these examples relate to bodily goal states: nourishment, keeping our muscles strong and our heart and lungs well functioning, physical and mental recuperation. The last relates to the need for mutual help, support, and encouragement, and to the many benefits arising from the exchange of ideas in conversation. The enlargement of our schemas that results from understanding is of very general pro-survival value because this increases the number of situations in which we can act appropriately. So it is no accident that we feel pleasure when we newly understand something. Here is a student's memory from early childhood.[2]

> When 'reading' the letters of the headline in the newspaper, I suddenly came to the realisation that the words were made up of sound groups I could recognise, which were represented as groups of letters—basically, I had made the discovery that the letters on the chart on my nursery wall were related to the words that I heard people speaking, and that I was attempting to articulate. I had discovered that both were part of the same idea. My mother reports that this discovery sent me racing round the house in ecstasy, attempting to 'pronounce' every written word I came across, and positively gurgling with pleasure.

**Unpleasure.**   This signals changes away from a goal state. We feel unpleasure when we miss our bus, when we lose our purse, or shiver in a cold wind. If at the same time there is nothing we can do about it, we also feel frustration, and all the examples of this kind contributed by my students have come into the latter category.

**Fear.**   This signals changes toward an anti-goal state: that is, one that is counter-survival. It warns us of danger. Thus, we feel fear when our car goes into a skid, or when we encounter a venomous snake. But survival is more than bodily survival, and that which threatens our self-image is also experienced as threatening.

> Fear has been the greatest emotion I have felt in a learning situation. Now, as I reflect on the situations in which I have been afraid, they are all related to a school

---

[2]This, and all the other extracts quoted in this and the next chapter, have been contributed by students taking my course *Foundations of Human Learning* at Warwick University. They have agreed to my quoting them, anonymously; and I am grateful to them for this valuable collection of examples.

environment. I have felt this fear as the form-positions are read out—the dread of coming lower than the position I should (in the teacher's opinion) be coming in.

**Relief.** This signals changes away from an anti-goal state. We feel relieved when the driver of the car we are in regains control after a skid. Again, the threat need not be a physical one.

I passed the eleven-plus exam[3] which was an enormous relief to me as I had been expected to. It is an awful amount of pressure for a child that age to be put under. I felt confident I would pass but then feared that would be asking fate to make me fail, so I used to walk home pretending that when the list of passes was read out in class, my name wasn't on it. I did this to practise not crying at the news as a girl had done the year before.

Relief is not the same as pleasure, and a poor substitute for it.

This cessation of finding pleasure in learning something in itself and instead experiencing relief, was also reflected in other areas of the classroom. The results of the tests were used as a basis for seating arrangements. The higher the mark, the further away from the teacher one sat. The unfortunate pupil who had got the lowest mark had the misfortune to sit right under the teacher's nose. Added to this was the shame and humiliation of everyone else knowing your marks.

The foregoing categories are summarised in Fig. 16.1.

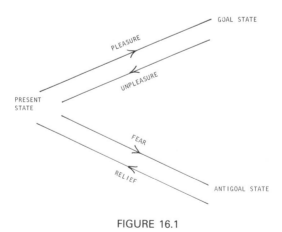

FIGURE 16.1

---

[3]The "eleven-plus" is an examination for admission to grammar schools, and other selective schools, that is still used in some parts of the United Kingdom.

## Emotions in Relation to Competence

The first four categories of emotion just described, relate to situations that have actually arisen, and call our attention to the need to do something. The next four relate to whether or not we are in fact able to do whatever is necessary to bring about the goal state, and to prevent the anti-goal state. They signal our own competence relative to the situation. This, too, has clear implications for survival because we need to be cautious about entering situations where we lack competence.

Confidence.    Confidence signals competence: ability to move toward a goal state. I sit down to write a letter on my word processor with confidence because I am well able to make it do what I want.

> English has always been my strongest subject, and so whenever I received a low mark, I felt able to cope with it because I knew I had the ability to do better, reflecting on past marks. A friend however, found the subject difficult and so each 'failure' she found harder to cope with.

Frustration.    This results from inability to move toward a goal state. When my monitor screen shows uninformative error messages like ''Error. Try again'' and I cannot discover the cause, I feel frustration.

> As long as I can remember I have been better at English than Mathematics: pleasure and confidence in the former and frustration and anxiety in the latter are an integral part of my school experience. Constant frustration in Mathematics has affected, as the model suggests, my ability in attempting certain new tasks. For example, unknown subject areas like computing (which is an important part of my psychology course). My belief in my inability in the Mathematical field has led to basic lack of confidence in an area where I know little, and have no previous experience.

Notice that we are talking about competence, the ability to achieve one's goals by one's own efforts. If it is a beautiful day, we may well feel pleasure; but we have no confidence in our ability to produce one of these when desired. The relation between these two emotions is well brought out by the next example.

> When I am playing a piece of music on the piano I experience pleasure. If I play a wrong note I feel unpleasure, but if I quickly correct myself pleasure returns. However, if when I play the wrong note, my brother leans over my shoulder and says 'F-sharp', I feel frustrated because I've been deprived of the chance of moving to the goal state by my own efforts.

Security.    Security signals that we are able to move away from anti-goals. A good climber will feel secure half way up a vertical rock face, not because there is no danger, but because he knows that he is in control of the situation.

> From my own experience it appears to be beneficial to do a number of questions related to a new concept in order to create security and confidence. . . . If one feels happy with what has just been learnt, there does not seem such a large risk involved in going onto the next stage.

**Anxiety.** If on the other hand we are in a situation where there are possible dangers, and we are unsure of our ability to avert these if they arise, then we are anxious. On an icy road, most drivers will feel anxious even when they are not actually in a skid.

This student is describing her feelings during reading lessons.

> As it neared my turn, I became more and more anxious as I was unable to do anything about it. . . . When it was my turn, anxiety was superceded by fear; fear that I wouldn't know all the words, that I'd lose my place or skip a line. I was fearful that the teacher would criticise me for not concentrating, but most fearful that my peers would laugh or think how stupid I was i.e. loss of peer-group status. When I was asked to stop reading, there was a feeling of immense relief.

## MIXED EMOTIONS

Because the same state may often be categorised in more than one way, it follows that more than one emotion may be aroused at the same time. These may be of similar or different kinds. Thus, a competent driver will feel confident that he can control the powerful engine of his car so that it will take him to his destination; and secure in his ability to avoid situations of danger, and to move away from them if they approach. He may also feel pleasure in the drive itself, toward a destination where he will enjoy a vacation. If his car develops a flat tyre, there will be frustration relative to the interruption of his journey, but also (provided he has a spare wheel with a sound tyre) confidence in his ability to regain his goal state of a well functioning car in which he and his passengers can continue their journey. This may be mixed with unpleasure at getting his hands dirty.

### Mixed Emotions in Learning

The foregoing were straightforward introductory examples. For our present purposes, we now need to apply these ideas to learning situations. These also may be classified in two ways: at the level of delta-one, which is the doing level; and at the level of delta-two, which is concerned with improving the abilities of delta-one (i.e., with learning). Thus we feel pleasure when we are aware that our ability is improving, unpleasure if we find we are getting worse.

> My father would very often shout, throwing me into a temper, however the pleasure in learning to drive came from my realisation that toleration of his temper would eventually mean that I could drive the car single-handed.

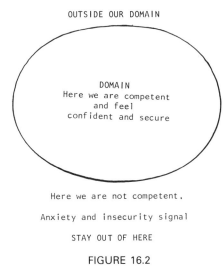

OUTSIDE OUR DOMAIN

DOMAIN
Here we are competent
and feel
confident and secure

Here we are not competent.

Anxiety and insecurity signal

STAY OUT OF HERE

FIGURE 16.2

It is in the nature of learning that this takes place in regions where we are not as yet competent.

In Fig. 16.2 the domain represents the region (set of states) where we can achieve our goals and avoid our anti-goals. This is our region of competence, and within it we feel confident and secure. In the region outside this domain we know that we are not competent. We can neither achieve our goals nor avoid our anti-goals, and feel both frustration and anxiety. These are strong signals to keep out of this region. Ancient maps sometimes warned "Here be monsters."

The boundary between inside and outside a person's domain is, however, usually not a sharp one (see Fig. 16.3).

There is a frontier zone in which we can achieve our goals, and avoid our anti-goals, sometimes but not reliably. It is in this area that learning takes place; and learning is thus a process of changing frontier zone to established domain. The frontier zone then moves outwards, and the process continues. It is pro-survival to expand our domains, so it fits in with this model that we have a strong exploratory urge. Because this takes us into our frontier zones, where we are not fully competent, the model also predicts that mixed emotions are to be expected in learning situations. These may be such as to bring learning to a halt, temporarily:

My brother is 6 years younger than me and I can remember very clearly when he first managed to take a few steps on his own. When he was 11 months old he tried to stand up on his own, but he fell down and he was really frightened. From that day he wouldn't take another single step. It was 5 months later . . . that my little brother was persuaded to try again. Fortunately, he managed this time, and although he could not talk, one could see how satisfied and pleased he was with

OUTSIDE OUR DOMAIN

FRONTIER ZONE

ESTABLISHED

DOMAIN

MIXED EMOTIONS

FIGURE 16.3

himself by the look on his face. He was obviously very happy and even his eyes were smiling.

Or even permanently:

I remember, a few years back, I used to envy my sister who could skate. Because it looked such fun, and because I didn't want to remain a spectator all the time, I decided to learn this skill. I was very unsure of the consequences and did not commence my learning with much, if any, confidence at all. . . . However, try I did, but no sooner had I set foot on ice then I found myself with a big crash on the floor, which was a very painful fall. Immediately this put me off, and I did not even attempt to have another go. Needless to say, I've never set foot in an ice-rink since.

Our first step has been to show that the model fits in with, and integrates, what we know empirically. The next step is to use the model to analyse the factors involved. This will suggest ways in which learning situations may be managed so that the negative emotional signals that may arise do not prevent learning from taking place. This, even more than giving information, I see as an important function of a teacher.

# Managing the Risks in Learning

The physical risks while learning may not be too difficult to control, by providing low-risk learning situations. Learner swimming pools are sometimes provided that are shallow enough for beginners to lie horizontally in a swimming position, off the bottom, but able to support themselves by putting their hands down. In this position they are in their frontier zone while their hands are off the bottom, but can easily regain full control by putting their hands down on the bottom again. A similar approach, for use in deeper water, teaches learners first of all how to regain a feet-on-bottom position from horizontal. In both of these, the principle is the same, and it is a general one: to give learners the confidence that they can regain their established domain whenever they want. This way, they feel much safer while in their frontier zones.

Today there are no longer physical risks in learning mathematics—although it is within living memory that children were caned for mistakes in calculation. But there are other ways, other than physical, of chastening, and these risks have not yet been taken away. Embarrassment, loss of self-esteem, feeling or looking stupid in front of other pupils, are often reported.

When I was about fourteen I asked my Mathematics teacher how to divide fractions. He was horrified and said in a loud voice so that all the class could hear, that he hoped I was joking. I cowardly said I was and to this day, even though I have read how to do it in Mathematics books, I feel convinced that I cannot do it. I was in the top Mathematics class at the time but had only scraped in, and that had made me feel more foolish at not knowing how to divide fractions. I didn't know because I had never been told.

Many teachers treat learning errors as misdemeanours, the result of lack of attention or of due effort. Proper management of learning risks requires that these be clearly distinguished. It is appropriate to treat genuine misbehaviour with disapproval: but task errors should be perceived as part of the learning situation, this being the nature of the path to competence. With this interpretation, errors are not something to be concealed, and ignorance is not something to hide in the hope that it will not be discovered. Rather, attention is focused on the error itself, and understanding of this error becomes a new sub-goal. By the process of understanding, the subject matter that began as error becomes assimilated to the schema in use, which is thereby enlarged. So this part of the frontier zone has been converted into established domain.

Here is an example that arose recently as I was writing this chapter. A student learning Logo had just begun to use recursion with change, and was experimenting with an attractive program called POLYSPI[1] which is rich in opportunities for exploration. She had used it to draw a variety of spirals that got bigger as drawing proceeded, and wanted to make one that got smaller. To do this, she gave a negative value to the variable: CHANGE. As she expected, the spiral began by getting smaller; but then to her surprise it started getting bigger again. Her exclamation of surprise brought me, and several fellow students, to see what was happening. From one point of view, she had a bug, which prevented her from achieving the goal she had met. But from a different point of view, she had arrived unexpectedly at a new and interesting result. The reason why it got larger was not obvious on the screen because the spiral she had produced was self-crossing. It was in fact also being drawn in the reverse direction. If you start with FORWARD 100, and change this by −5 each recursion, the lines will indeed get shorter for a while, but will then start getting longer in the opposite direction.

The point of this example is the process of re-classification of the bug. If the main goal had been to draw a particular shape, the bug would have been perceived as an obstacle, and the dominant sub-goal would have been to eliminate it. In fact her main goal was not this 'doing' goal, but the learning goal of learning to use recursion with change, for which this program was just one example. When she encountered the bug, understanding the bug itself became, for the time being, the dominant goal. (This attitude to the treatment of bugs closely matches with that advocated by Papert, 1980, as do our educational philosophies.) This was more important for progress along the learning path than making that particular program work. The step forward along this path that resulted from understanding the bug was experienced by all of us as a shared pleasure.[2]

---

[1]To be found in the *Logotron Logo Handbook* (1984). Paris: Systemes d'Ordinateurs Logo International.

[2]With these students there was also a superordinate goal, that of reflecting on and discussing their own learning processes.

From the foregoing, we may summarise our first principle of risk management.

(i) Errors should be treated as potential contributions to learning: certainly not as misdemeanours.

Another factor that influences which way the emotional balance tips is the relative importance attached to learning goals and doing goals. This is related to the order in which one tries to achieve them. If expansion of one's competence is the dominant goal, failure at a particular task is less important. Thus, when I have bought a new utility for my word processor, I don't immediately try to use it for my work. This would be likely to cause frustration, and also divided attention because I would be trying to think both how to make the new utility work and about the chapter I was working on. The goal switches from putting thoughts into words to achieving a new state of competence in physically managing the words themselves. I might well say "I'm going to play with my new utility," because this is how it feels. And this is what, in the present context, I mean by play: a low-risk learning situation. Unlike most of the learning I did at school, *this is fun*.

As my competence develops, I do begin to use the new utility for my work. This gives pleasure of another kind because now I have a new ability to reach goals that matter, more easily and reliably. By doing it this way round, I have experienced only pleasure throughout.

I was able to do it this way because I had insight into my own learning processes, and a high degree of personal freedom in managing them. Both of these are commonly denied to school children. If (as I do not accept) this has to be so, then the least we can do for them in return is to help them to understand more about their own learning processes, and to manage the situation in ways that reduce the emotional risks. The foregoing discussion has added three more to our principles of management. These are:

(ii) Separate learning goals from doing goals;
(iii) set learning goals first; and
(iv) as competence increases, use doing goals for consolidation, and to contribute pleasure at the doing level also.

I can hear someone responding "But we learn by doing." In some subjects, such as the acquisition of a physical skill, yes. Even so, it is important first to analyse and have clearly in mind the details of the skill we want to learn, or we may learn bad habits that are hard to eradicate. This is part of the job of the golf coach, the teacher of singing or the violin. In the case of mathematics, however, what we need to learn are the schemas that are the sources of action. This route, although indirect, is ultimately more powerful. It is the path of research in the natural sciences also; and I believe that the research scientist in his laboratory, and the child at *play* (as I am here using the term) have more in common than either of them have with children of compulsory school age working for examinations. Any reader who deduces that I am suggesting a more playful approach to the learning of mathematics will be quite right.

## CONFIDENCE AND FRUSTRATION TOLERANCE

Frustration is another likely concomitant of a learning situation, if the goal is not yet within one's capabilities. Future success may then depend on how an individual deals with the emotions that are aroused.

Some are able to find resources of hope which enable them to view a problem constructively, while others are overcome by feelings of frustration or helplessness. I can illustrate this observation with my experiences as a student teacher on teaching practice in a local primary school. A lesson plan was drawn up involving the use of language and sentence structure. The children were Asian and, as English was their second language, they were unable to cope with the exercises set for them. One child 'G' remained calm and asked for help and spelling corrections, not understanding any better than the other children, but nevertheless, willing to try. Another child 'R' immediately said "I don't care if I can't do it" and gave up, speaking on the defensive to hide his lack of confidence in his own ability.

In Bruner's expressive terms, the first child was coping with the situation, the second was defending. Long term, the difference has far-reaching consequences. A person who has confidence in his ability to learn will be able to tolerate for a longer time the frustrations that arise while in his frontier zone. The longer he sticks at the learning task, the greater the likelihood of success; and vice versa. So confidence, or its lack, are likely to be self-fulfilling prophesies. Experience of success in learning builds up confidence in ability to learn, and increases the likelihood of staying on task long enough to do so. So a high priority is to provide a learning environment that is supportive.

In the course of my education, I had one teacher in particular who strove to provide this emotional support. He tried to create as relaxed an atmosphere as possible, and tried to encourage everyone to make some contribution, no matter how small, without actually putting pressure on anyone to do so. . . . Thus, emotional support was provided both by the teacher and by the other pupils. This was of great importance and benefit when we were learning new concepts, as the risk from being in our frontier zones did not seem so great. In situations where little or no progress was made towards a goal state, compensatory emotions were provided in the absence of pleasure that resulted from this.

In contrast:

> Each lesson was a torment, with the curriculum being sped through and me (and a few others) being left behind. Mr C was a very frightening man, and to ask him for help was a feat in itself. But even then, if I did not understand his explanation of a point, instead of explaining it again slowly or in a different way, he seemed to just shout louder and thump his hand on the desk in emphasis of certain points.

A threatening learning environment does not make for intelligent learning.

> The stress of the situation invariably diminished one's ability to concentrate and think intelligently. Even if you were not the 'victim' at that precise moment, there was still the possibility that you would be next.

So our next principle of risk management is one that I would regard as almost too obvious to state, were it not for the frequency of counter-examples such as those just given.

(v) Maintain a learning environment that is supportive and aids the growth of confidence, not one that is threatening and destructive of confidence.

Another factor that helps the growth of confidence is that of freedom to enter or leave the frontier zone according to one's own needs. Two bright seven-year-olds were working (or should I say playing?) at Logo. For some time they had been doing what was for them quite easy; so I asked if they would like to learn a new command. They said that they would, so I introduced them to REPEAT, which they liked and used with success to draw a square. Then they went back to what they had been doing before. I was a little surprised at this, and wondered if I had intervened too soon. (My approach to teaching LOGO is that of minimal, but not zero, intervention; and my own learning goal in this situation was to develop criteria for the nature and timing of these interventions.) Reflection, in terms of the present model, reassured me. They had been ready to move into their frontier zone for a while, and had enjoyed learning something new. But afterwards, they chose to return for a while to the security of their established domain. The learning situation I provided allowed them to do so. In this way, any feelings of frustration and/or anxiety are allowed to subside. By working for a while within a learners' area of competence, confidence is regained, and they may then feel ready to return again to their frontier zone for further learning.

This contrasts with the common practice of relentlessly setting a new learning goal as soon as the current one has been achieved. So the next two principles of risk management are:

(vi) Allow time for consolidation of a new frontier zone before pushing on into the unknown.

(vii) So far as is practical, allow learners to choose the pace at which they expand their frontiers of knowledge.

## Panic

In some cases, the emotional stress is such that students have put an even stronger name to it.

> Friday morning was Mathematics, which for me was an awful experience. I panicked and could never get my sums finished and correct, and my teacher would not allow me to join in hobbies until I got at least one correct answer. Mathematics then meant failure, punishment and ultimately humiliation.

Buxton has made an extensive investigation into this specific issue of panic in mathematical situations, using a variety of methods. These included individual interviews in depth; group learning situations with voluntary adult subjects, continuing over a period; and single experiments with groups. He describes one of the latter as follows (Buxton, 1985; see also Buxton, 1981).

> I opened the proceedings with a statement of this sort: "I am Staff Inspector for Mathematics and I am going to give you a test. I shall want to see your answers so that I can judge them. There is a strict time limit." (Takes off watch and holds it.)
>
> After a slight pause, I would then say: "Now I want you to close your eyes, find the single emotional word that best describes your feelings and write it down. Do not put your name on the paper."
>
> The release of tension was usually very marked, and accompanied by gusts of nervous laughter. When the results had been collected, I read them back to the audience. There was nearly always a proportion in the panic region.

Here is a typical list of responses from one of these experiments, in descending order of severity. Terror, panic, panic, panic, sweaty/palpitating, fear, fear, Ugh!, apprehension, apprehension, tense. tension, tension, nervous, hilarity, ridiculous, stupid, joy.

Buxton (1985) analyses situations such as these as follows.

The major problem arises when the threat is seen as serious enough not to want to suffer the consequences of ignoring it, when delta-one does not have a routine, and when delta-two fails. It may fail:

(a) because it cannot make a plan.
(b) because it cannot make a plan *in time*.
(c) because it has *no expectation* of being able to make a plan.

> The threat directs the consciousness into delta-one, which then refers it to delta-two. [Because it does not have a routine, it needs a new plan to be made.] When a plan does not result, the emotions generated through the comparator again bring the consciousness to delta-one, asking that it deal with the threat, but again it is

referred to delta-two. The process then becomes a flicker of the consciousness between the two deltas; the result is paralysis and a state of complete disfunction. (pp. 53–54)

Buxton was able, reliably, to cause emotions such as those just mentioned in mature adults by imposing two pressures: of positional authority, and of time. School teachers are likewise in positional authority over their pupils, and if one of them has been asked a question and everyone is waiting for a reply the second pressure is also present. So the resemblance between the feelings described by Buxton's subjects, and by the students whose reports have been cited earlier, matches the similarity of their situations.

Buxton was also able to change the emotions felt by his same subjects in a mathematical situation into ones that were almost entirely favourable. The four who initially reported terror or panic now reported, respectively: "Much more relaxed. Interested." "Quiet fun, which I really enjoyed." "Happy, calm and relaxed." "Calm, more secure but still had moments of feeling not confident." The two main methods that he used were first, to say that he was not going to see their answers, giving as reason that "the authority lies within the subject (maths.)"— and when they have found it there will be no need to check with my (sapiential) authority. And second, to take away the time pressure. ". . . the fact that there is no time limit must be stated, often several times."

Although I think that it may sometimes be useful for a teacher to see or hear learner's responses, the emphases on sapiential authority, and the authority of the subject, are important. They are different aspects of mode 3 schema testing, consistency with the accepted body of knowledge; and sapiential authority is exercised by helping a learner to see where his ideas do, or do not, fit in with what he already knows. The goal is understanding, rather than obedience. This distinction between the two kinds of authority, structural and sapiential—authority of position and authority of knowledge—I regard as very important, particularly since as many teachers find themselves situated there is role conflict and role confusion. I have discussed it at some length elsewhere (Skemp, 1979a, Chapter 15).

From the foregoing, two further principles of risk management may be summarised.

(vi) Distinguish clearly between authority of position and authority of knowledge.

(vii) Allow plenty of time for reflective intelligence to function.

## CONFIDENCE AND UNDERSTANDING

We must never forget the contribution that understanding makes to confidence in a new situation.

At school in my maths class, when the teacher wrote something on the board, often my mind would go blank. Even though he'd give a thorough explanation as to what it was, and how it got there, I would still have great difficulty in understanding it. I'd copy down everything he'd say, and hope that he would not ask me any questions, for I would not be able to answer. When at home, I'd start to read over what he'd written, and try to build up a mental picture of what it meant. If the new idea didn't click, the homework would take a long time, for I'd have to follow his examples closely, and copy the way he had done it. If I got the first few right, my confidence would increase, I could relax a little, and the questions would get slightly easier to answer. If they weren't correct, I'd start to worry, because I'd hate to go back to school the next day and not be able to answer a question in front of my classmates. If the idea did click, then homework would be easier, and school would again be a 'safer' place to be, because I knew I could cope. If the new ideas had fitted into a regular pattern/schema, then they would be easy to remember as well. Things I memorized, soon needed rememorizing, whereas concepts that had clicked just seemed to be there. Some short revision, could quite easily bring them back into consciousness.

This student was anxious when confronted with new work to be learned. Intuitively, however, he knew the difference between habit learning (memorizing) and learning with understanding (ideas that did click). He also knew the greater ease of remembering that which he had understood. ("Things I memorized, soon needed rememorizing, whereas concepts that had clicked just seemed to be there.") As a result, he set himself the right learning goal, and when he achieved this his worry changed to security ("school would again be a 'safer' place to be, because I knew I could cope").

With habit learning, there is little frontier zone, if any. Any situation that is different from those for which the learner has memorised rules throws him back on the teacher for new rules, or for examples that provide a method that can be copied. With intelligent learning, however, the frontier zone is such that it can be assimilated to an existing schema. One's cognitive map leads part way into the frontier zone. Its features have about them something familiar, so that they can be understood by expanding and/or extrapolating existing ideas. For example, a child who can count in units and tens is able to see the same pattern repeated in hundreds and thousands. By extrapolating this pattern in the reverse direction, place-value notation (IF well understood) can help him to understand tenths and hundredths.

Thus, an important contributor to overall confidence within the frontier zone is confidence that one can understand the new ideas—can assimilate them to the well understood schema that constitutes his established domain. So the two principles described in Chapter 2 (page 18), and their applications as there described, are important in the present context also. It would be good to review these at the present stage. To add them to our list of principles for risk management, I summarise them here.

(viii) By a process of conceptual analysis, tested by teaching experiments, the teacher must have available a dependency network, or concept map, showing which concepts are pre-requisite for others. The learning path must then be planned and guided so that when entering his frontier zone, a learner always has available the necessary schemas to give him confidence in his ability to understand the new material.

## PEER-GROUP INTERACTION IN LEARNING

A learning situation that has been found to provide a good emotional climate for intelligent learning has been developed over a number of years while I was working in primary schools in England and Wales (Skemp, 1985). Among the features emphasised is the value of peer-group interaction in a learning situation.

A class organisation in which children work together in small groups is already well established in many primary schools. In mathematics, however, this is seldom put to use: children still work individually from textbooks and work cards, which they take to their teacher for correction.

As an alternative for this approach, we have provided a carefully structured collection of mathematical activities for use by children in small groups. Some of these involve co-operation in learning; others are games based on mathematical ideas, in which success largely depends on mathematical thinking. Incorrect moves are likely to be challenged by other players on mathematical grounds. In this way, we provide a shared mathematical experience that gives rise to discussion, mutual help, and explanation. Although many of the games are competitive, this is light-hearted and within an overall situation of co-operation. I say, ''When we're learning, it is good if we help each other''; and it is my experience that children do this in learning situation of the kind I have described.

This approach was introduced initially in order to make fuller use of all three modes of schema construction (see Table 8.1, page 110). The good emotional climate that results, of mutual help and shared enjoyment in learning, has been an important additional factor of the success of this new approach to the intelligent learning of mathematics.

# The Silent Music
# of Mathematics[1]

The thoughts which follow result from a combination of two events which took place last Christmas. One of these was a visit from a niece of mine who has two bright children aged seven and eight. She was worried because all the mathematics they did at school was pages of 'sums.' Shortly before this, we had heard and seen a performance of Benjamin Britten's beautiful *Ceremony of Carols*. This was introduced by Britten's lifelong friend Peter Pears, who related how it had been composed by Britten at sea, in a cramped cabin with no piano or other musical instrument on which Britten could hear what he was composing. Afterwards, I began to wonder how such music could be composed under these conditions. How could he know how wonderful it would sound in performance? Maybe he sang to himself, some of it. But he could only sing one part at a time, and what about the harp? My answer cannot be more than a conjecture; but I think we may assume that like many other composers he was able to write music directly from his head onto paper because he could hear the music in his mind. The musical notation represented for him patterns of sound, sequential and simultaneous.

There are others besides composers who can hear music in their minds, and who (we are told) can get pleasure from reading a score in the same way as others enjoy reading a book. But most of us are not like this. We need to hear music performed, better still to sing or play it ourselves, alone or with others, before we can appreciate it.

---

[1]Reprinted from *Mathematics Teaching*, no. 102, March 1983, p. 58.

We would not think it sensible to teach music as a pencil and paper exercise, in which children are taught to put marks on paper according to certain rules of musical notation, without ever performing music, or interacting with others in making music together. To start with, children are not taught to read or write music at all—they sing, listen, and move their bodies to the sounds of music. And when they do learn musical notation, it is closely linked with the performance of music, using their voices, or instruments on which they can play without too much difficulty.

If we were to teach children music the way we teach mathematics, we would only succeed in putting most of them off for life. It is by hearing musical notes, melodies, harmonies and rhythms that even the most musical are able to reach the stage of reading and writing music silently in their minds.

So why are children still taught mathematics as a pencil and paper exercise which is usually somewhat solitary? For most of us mathematics, like music, needs to be expressed in physical actions and human interactions before its symbols can evoke the silent patterns of mathematical ideas (like musical notes), simultaneous relationships (like harmonies) and expositions or proofs (like melodies).

Regretfully, I hold Mathematicians (with a capital M) largely to blame for this. They are so good at making silent mathematics on paper for themselves and each other that they have put this about as what mathematics is supposed to be like for everyone.

We are all the losers. Music is something which nearly everyone enjoys hearing at a pop, middle-brow or classical level. Those who feel they would like to learn to perform it are not frightened to have a go, and those who perform it well in any of these varieties are sure of appreciative audiences. But Mathematicians have only minority audiences, consisting mostly or perhaps entirely of other Mathematicians. The majority have been turned off it in childhood. For these, the music of mathematics will always be altogether silent.

# References

Allardice, B. B. (1977). The development of written representations for some mathematical concepts. *Journal of Children's Mathematical Behaviour, 1,* 4.

Ausubel, D. P., & Robinson, F. G. (1969). *School learning: An introduction to educational psychology.* New York: Holt, Rinehart & Winston.

Backhouse, J. K. (1978). Understanding school mathematics—A comment. *Mathematics Teaching,* No. 82, pp. 39–41.

Bartlett, F. (1932). *Remembering.* Cambridge, England: Cambridge University Press.

Behr, M. J., & Post, T. R. (1981). The effect of visual perceptual distractors on children's logical-mathematical thinking in rational-number situations. In T. R. Post & M. P. Roberts (Eds.), *Proceedings of the Third Annual Meeting of the North American Chapter of the International Group for the Psychology of Mathematics Education* (pp. 8–16). Minneapolis: University of Minnesota.

Bell, E. T. (1937). *Men of Mathematics.* Chap. 19. Harmondsworth: Penguin Books.

Bell, M. A. (1967). Unpublished M. Sc. thesis, University of Manchester.

Bondi, H. (1976). The dangers of rejecting mathematics. *Times Higher Education Supplement,* 22 April 1976, pp. 8, 9.

Bruner, J. S. (1960). *The process of education.* Cambridge, MA: Harvard University Press.

Buxton, L. (1978a, July). What goes on in the mind? *Times Educational Supplement,* 18–19.

Buxton, L. (1978b). Four levels of understanding. *Mathematics in School, 7*(5), 86.

Buxton, L. G. (1981). *Do you panic about maths?* London: Heinemann Educational.

Buxton, L. G. (1985). *Cognitive-affective interaction in foundations of human learning.* Unpublished doctoral thesis, University of Warwick.

Byers, V., & Herscovics, N. (1977). Understanding school mathematics. *Mathematics Teaching,* No. 81, pp. 24–27.

Carpenter, T. P. (1979). *Cognitive development research and mathematics education.* Center theoretical paper no. 73, Wisconsin Research and Development Center for Individualised Schooling, Madison, WI.

Cockcroft, W. H. (1982). *Mathematics counts.* London: Her Majesty's Stationery Office.

Davis, R. B. (1984). *Learning mathematics: The cognitive science approach to mathematics.* London & Sydney: Croom Helm.

Dewey, J. (1929). *Sources of a science of education.* New York: Liveright.

Erikson, E. E. (1950). *Childhood and society.* Harmondsworth: Penguin.

Ghiselin, B. (Ed.). (1952). *The creative process.* Berkeley: University of California Press.

Ginsberg, H. (1977). *Children's arithmetic: The learning process.* New York: Van Nostrand.

Glennon, V. J. (1980). *Neuropsychology and the Instructional Psychology of Mathematics.* Kent, OH: Research Council for Diagnostic and Prescriptive Mathematics.

Hadamard, J. (1945). *The Psychology of Invention in the Mathematical Field,* New York: Dover.

Hart, K. M., Brown, M. L., Küchemann, D. E., Kerslake, D., Ruddock, G., & McCartney, M. (1981). *Children's understanding of mathematics: 11–16.* London: Murray.

Hirabayashi, I. (1984). Critical period to prepare pupils for the secondary school mathematics. Adelaide, Australia: Short communication, ICME V.

Holt, J. (1969). *How children fail.* Harmondsworth, Penguin.

Inhelder, B., & Piaget, J. (1958). *The Growth of Logical Thinking,* London: Routledge and Kegan Paul.

Kieran, C. (1979). Children's operational thinking within the context of bracketing and the order of operations. In D. O. Tall (Ed.), *Proceedings of the Third Annual Meeting of the International Group for the Psychology of Learning Mathematics* (pp. 128–133). Coventry: Mathematics Education Research Centre, University of Warwick.

Kuhn, T. S. (1970). *The structure of scientific revolutions.* Chicago: University of Chicago Press.

Lawler, R. W. (1982). Designing computer-based microworlds. *Byte, 7*(8), pp. 138–160.

Macfarlane Smith, I. (1964). *Spatial Ability,* London: University of London Press.

Morris, D. (1967). *The naked ape.* London: Cape.

The National Commission on Excellence in Education. (1983, April). *A nation at risk: The imperative for educational reform.*

Opper, S. (1977). Piaget's clinical method. *Journal of Children's Mathematical Behaviour, 1* (4), 90–107.

Papert, S. (1980). *Mindstorms.* Brighton: Harvester.

Piaget, J. (1928). *Judgement and Reasoning in the Child,* London: Routledge and Kegan Paul.

Popper, K. (1976. *Unended quest: An intellectual autobiography.* Glasgow: Fontana/Collins.

Sartre, J.-P. (1964). *Les Mots.* Paris: Gallimond.

Skemp, R. R. (1961). Reflective intelligence and mathematics. *British Journal of Educational Psychology, XXXI,* 45–55.

Skemp, R. R. (1962). The need for a schematic learning theory. *British Journal of Educational Psychology, xxxii,* 133–142.

Skemp, R. R. (1962–1969). *Understanding Mathematics* (7 volumes). London: University of London Press.

Skemp, R. R. (1971). *The psychology of learning mathematics* (1st ed.). Harmondsworth: Penguin.

Skemp, R. R. (1976). Relational understanding and instrumental understanding. *Mathematics Teaching,* No. 77, pp. 20–26.

Skemp, R. R. (1979a). *Intelligence, learning, and action.* Chicester & New York: Wiley.

Skemp, R. R. (1979b). Goals of learning and qualities of understanding. *Mathematics Teaching,* No. 88, pp. 44–49.

Skemp, R. R. (1983). The functioning of intelligence and the understanding of mathematics. In M. Zweng, T. Green, J. Kilpatrick, H. Pollak, M. Suydam (Eds.), *Proceedings of the Fourth International Congress on Mathematical Education* (pp. 532–537). Boston: Birkhouser.

Skemp, R. R. (1983). The school as a learning environment for teachers. In J. C. Bergeron & N. Herscovics (Eds.), *Proceedings of the Fifth Meeting of the North American Chapter of the International Group for the Psychology of Mathematics Education* (pp. 215–222). Montreal: Université de Montréal and Concordia University.

Skemp, R. R. (1985). PMP (Primary mathematics project for the intelligent learning of mathematics): A progress report. In L. Streefland (Ed.), *Proceedings of the Ninth International Conference for the Psychology of Mathematics Education* (pp. 447–452). Utrecht: State University of Utrecht.

Steffe, L. P. (1977, Fall). *The teaching experiment.* Unpublished manuscript presented at a meeting of the models working group of the Georgia Center for the Study of Learning and Teaching Mathematics, University of New Hampshire.

Steffe, L. P., Richards, J., & von Glasersfeld, E. (1979). Experimental models for the child's acquisition of counting and of addition and subtraction. In W. Geeslin (Ed.), *Explorations in the modeling of the learning of mathematics* (pp. 29–31). Columbus, OH: ERIC/SMEAC.

Tall, D. O. (1977). Conflicts and catastrophes in the learning of mathematics. *Mathematical Education for Teaching, 2*(4), 2–18.

Tall, D. O. (1978). The dynamics of understanding mathematics. *Mathematics Teaching,* no. 84, pp. 50–52.

Vygotsky, L. S. (trans. Hanfmann, E., & Vakar, G.) (1962). *Thought and Language,* Cambridge, MA: M.I.T. Press.

Weizenbaum, J. (1976). *Computer power and human reason.* San Francisco, CA: Freeman.

Whitney, H. (1985). Taking responsibility in school mathematics education. In L. Streefland (Ed.), *Proceedings of the Ninth International Conference for the Psychology of Mathematics Education* (Vol. 2, pp. 123–141). Utrecht: State University of Utrecht.

# Author Index

# Subject Index